Finite Groups and Quantum Theory

Finite Groups and Quantum Theory

D. B. CHESNUT

Professor of Chemistry
Duke University
Durham, North Carolina

A Wiley-Interscience Publication
JOHN WILEY & SONS
New York · London · Sydney · Toronto

Library of Congress Cataloging in Publication Data:

Chesnut, D B 1932–
Finite groups and quantum theory.

"A Wiley-Interscience publication."
Bibliography: p.
1. Quantum theory. 2. Finite groups. I. Title.

QC174.17.G7C47 530.1′2 74-9840
ISBN O-471-15445-8

Printed in the United States of America

10 9 8 7 6 5 4 3 2 1

to Debbie, Lauren, Blair, and Lynn

Preface

Group theory has been described as organized common sense. To this should be added that the sense it makes often depends on the organized form in which it is presented. It has found its way into many introductory texts on quantum chemistry and is present in the curricula of the good physical chemistry program. The defining properties of a group are readily and quickly stated, and in a matter of three or so lectures some elementary skill in the use of the theory in certain problems can be imparted. Although this approach may appeal to the scientific pragmatist, much of the sweeping power, beauty, and outright fun of the subject are lost. At the opposite extreme are the highly mathematical treatments that tend to make little connection with the physical world and quickly attenuate the interest of the physical student. A proper middle ground should be one in which the student gains competence and confidence in, as well as an appreciation of, the use of the theory in physical problems and at the same time receives an adequate education in the basic mathematical background involved. Much of the basic math involved is straightforward and is easily assimilated once the proper definitions and terminology are introduced.

The aim of the book is to present in a readable way the basic mathematical ideas and structure behind the theory of point groups, emphasizing its basic use and application in quantum mechanics. It does not pretend to be complete but does venture into some areas normally omitted in the briefer presentations of the topic at the beginning level. The reader will learn how to use and apply the theory to basic quantum mechanical problems while gaining knowledge of the related mathematical background and terminology.

The book is written for beginning and intermediate graduate students, although it should not be beyond seniors in good undergraduate programs. The reader is expected to have had some exposure to the elements of quantum theory and should be competent in matrix theory and the terminology and classification of point group symmetry. Based on the author's experience and background and the examples of applications taken up in this book, chemistry and chemical physics students will feel at home.

One should try to write as he tries to lecture—with an eye to the student's understanding. "Understanding" really means "feeling comfortable with," and, as with any subject, this will only occur with practice and study. To aid in these matters, problems and exercises have been included in the body of the text, and *all* the solutions are to be found at the end of the appropriate chapter. Some of the problems will test the reader's understanding of material already illustrated by example in the text; other problems present theorems, proofs, or properties not explicitly discussed in the text but which the reader should be able to attack. The presence of solutions is particularly important for the latter category of exercises since the combined problem and solution will read as an extension of the text having first allowed the reader a crack at strengthening his skills by attempting to derive it himself. One only learns a subject well by practice, and the separation of the problem solutions from the problems by a few pages is an attempt to help the reader resist the urge to turn to the answer before working it himself. I think it important, however, that the solutions be present on demand, for the time to learn is the time when one is interested. If one has to wait several hours or several days to see someone else's solution, the impact of the problem and its educational value may well have been lost.

Group theory divides itself rather naturally into the three basic areas of abstract theory, representations, and applications. This is the division found in most texts and is that used here. Before defining a group, the basic concepts in set theory are presented; the idea of groups as sets with algebraic structure is then introduced by the special sets important in the partitioning of groups such as cosets, factor groups, and classes. Mapping—the bridge between the abstract theory and its nonabstract representations—is briefly discussed. Because of its general importance in the physical sciences a chapter is included on permutations and the permutation groups. The section on representations begins with a general examination of linear transformations which not only sets the stage for the construction of group representations but also reviews the student's background in matrices, vectors, and their manipulation. Sections follow on irreducible representations, the character, the orthogonality relations, character tables, and direct product representations. The basic and fundamental ideas of quantum theory are reviewed, introducing certain ideas and terminology that mesh well with the mathematics of group theory. The symmetry of the Hamiltonian and the irreducibility postulate follow along with a discussion of operators, especially projection operators. The construction of symmetry-adapted basis functions is illustrated through examples and exercises. The applications of the theory are treated in the final three chapters, which delve into matrix elements and symmetry considerations, the permutation symmetry of wave functions, and the use of group theory in orbital theories of quantum mechanics.

Both the theory and the applications tend toward the mathematical side with an emphasis on the fundamentals involved. Many other more detailed applications of the theory can be found elsewhere, and the reader should find the presentation there much more understandable after the treatment of the fundamentals given in this book. There are, of course, many other aspects of group theory, particularly those dealing with continuous groups, which have not been treated in this book. The primary goal has been the treatment of the theory of point groups; the treatment of rotation groups, double groups, time inversion, and many other interesting topics must await an expanded version of this book or the reader's own investigation into the more detailed and advanced treatments of the subjects.

The notation employed is for the most part standard. Schönflies point group symbols are used along with the generally accepted group theoretical symbols. To keep their role clear, operators are indicated by a superscript caret. Matrices are written in capital boldface and vectors, usually, in lowercase boldface. Initially, some "classical" vector notation involving arrows is used for emphasis and to ease the transition from ordinary Cartesian vectors to the more general n-component vectors.

Although the aim of the book has been one that the interested student can handle on his own, it can, of course, be the basis of a course in the subject. Perhaps such a book really is best used in conjunction with a formal learning situation because it lends the lecturer greater flexibility in what he himself will discuss and emphasize, knowing that the basic facts are available to the student. The material in this book has served as lecture material for a one semester course. Having the material in print should allow for perhaps a shorter time of coverage of the basic ideas, so that a one semester coverage with additional work, practice, and outside reading would represent the optimum treatment.

I am indebted to numerous people for the fact that this book exists: to my own teachers and those authors before me who instilled in me an appreciation of the usefulness and beauty of group theory; to Professors William Rhodes and Michael Kasha and the members of their research groups for their kind hospitality to me during a semester's sabbatical stay at Florida State University where the initial writing on this book was begun; to Sam Ferguson and Allan Butterfield for reading parts of the manuscript for clarity and error; to Mrs. Rachel Salter who cheerfully typed the many drafts and final copy of this manuscript in a superb professional fashion; and to my family who encouraged me in this endeavor and who give meaning to all my work.

If you as reader or teacher find this presentation of group theory enjoyable and fun to digest and employ, and if you have gained some appreciation for the scope and power of the theory, the success of the book in the eyes of the author will be assured.

D. B. Chesnut

Durham, North Carolina
April, 1974

Contents

1 SETS 1

2 GROUPS. 9

Definition of a Group 9
Multiplication Tables 10
Subgroups 12

3 THE PARTITIONING AND FACTORING OF GROUPS . . 19

Cosets and Lagrange's Theorem 19
Factor Groups 22
Products of Groups 23
Conjugate Elements and Classes 26

4 MAPPING AND GROUPS 35

The Mapping of Sets 35
The Mapping of Groups 37
Factor Groups and Homomorphisms; the Isomorphism Theorem 39

5 PERMUTATION GROUPS. 43

Permutations 43
The Symmetric Group. 45
Cayley's Theorem 46
The Alternating Group 50
Class Structure and Permutations 51

6 VECTORS, MATRICES, AND TRANSFORMATIONS: A REVIEW. 57

7 REPRESENTATIONS AND THEIR CONSTRUCTION . . . 63

The Definition of Representations 63
Linear Transformations 65
The Regular Representation 70
Related Representations 71
The Addition of Representations 72
Direct-Product Representations 74

8 IRREDUCIBLE REPRESENTATIONS **85**
Criteria for Irreducibility. 85
Some Basic Theorems 88
The Orthogonality Relations 89

9 THE CHARACTER **95**
Definition and Basic Properties. 95
Character Orthogonality Relations 96
Character Tables 99
The Decomposition of Reducible Representations 101
The Regular Representation 104
Abelian and Cyclic Groups; Complex Conjugate Characters 106

10 THE TRANSFORMATION OF FUNCTIONS AND COORDINATES **115**
The Transformation of Coordinates. 115
The Transformation of Functions. 118
Substitution Operators. 121
Examples of Function Transformations 121

11 PROJECTION OPERATORS AND SYMMETRY ADAPTED FUNCTIONS **129**
General Projection Operators 129
Group Projection Operators 130
The Use of Symmetry Adapted Functions. 137

12 GROUP THEORY AND QUANTUM MECHANICS. **145**
Quantum Theory 145
A Digression on Operators and Operator Representatives. . 147
The Symmetry Group of the Hamiltonian 151
Symmetry and Energy Eigenfunctions 153
The Irreducibility Postulate 155
Accidental Degeneracy 156
Conclusion 159

13 MATRIX ELEMENTS AND SYMMETRY CONSIDERATIONS. **167**
The General Case. 167
Decomposition of the Direct Product; Independent Functions 170
A Specific Situation 171
The Case of Dependent Product Functions 172
The Symmetrized and Antisymmetrized Product 173
Summary 175

14 THE PERMUTATION SYMMETRY OF WAVE FUNCTIONS 181

Wave Functions and Orbitals 181
The Permutation Symmetry of Physical Systems 183
Fermions and the Exclusion Principle 186
Matrix Elements and Antisymmetrized Orbital Products . . 187

15 ORBITAL APPROACHES TO CHEMICAL BONDING . . . 199

Molecular Orbital Theory 199
Valence Bond Theory and Directed Valence; Hybridization . 206

APPENDICES 225

A. Some Aspects of Class Algebra 225
B. Symmetry Reducing Perturbations and Group Correlation Tables 231
C. Selected Character Tables 235
D. Reading List 249

INDEX 251

Finite Groups and Quantum Theory

1
Sets

A group is a set with algebraic structure whose composition law satisfies certain postulates. To appreciate this basic definition of a group we clearly must know something about sets. The concepts involved with sets are useful not only in their own right but also because they facilitate our understanding of many of the concepts in group theory.* In addition, certain notation we will use in our later development will be introduced here.

A *set* is simply a specified collection of objects or things; the objects or things making up the set are called its *elements* or *members*. A set may be defined without specifying the mathematical nature of its elements. We can have a set of students in a classroom (a finite set), or the set of all positive integers (an infinite set); "numbers" represent that property of sets associated with how many elements there are in particular sets. If we use capital letters for sets and lower case letters for elements, we specify a set as, for example,

$$A = \{a, b, c\} \tag{1.1}$$

and denote that "a" is an element of A or is contained in A by

$$a \in A \tag{1.2}$$

where the symbol "$\in$" is translated "belongs to" or "is a member of." The order of the elements inside the curly brackets is not important, although often a particular order may be convenient for characterizing the nature or structure of the set in question. If an element "d" does not belong to A, we write

$$d \notin A \tag{1.3}$$

It is presumed that the elements of a set are distinctive; that is, they are different from each other in some respect.

* A clear and concise discussion of the use of set concepts and symbolism in the mathematical logical sense is given in W. Feller, *An Introduction to Probability Theory and Its Application*, 3rd ed., Wiley, New York, 1968, Chapter 1.

We can also specify membership in a set by defining the characteristic(s) of its elements. Thus, for example,

$$A = \{a;\ a \text{ is an even integer greater than 0 and less than 10}\}$$

or, as before

$$A = \{2, 4, 6, 8\} \tag{1.4}$$

Just as we can have an infinite set (a set with an infinite number of elements) so too can we have a set with no elements. This set, obviously unique, is called the *null* set and is usually denoted by the symbol $\varnothing$. The null set in set theory plays a role similar to that of the integer zero in number theory.

There can be smaller collections of elements within a larger collection; these are also properly called sets and are *subsets* of the larger set. If B is a subset of A we so denote this by

$$B \subset A \tag{1.5}$$

In the above example, if

$$\begin{aligned} B &= \{2, 4\} \\ A &= \{2, 4, 6, 8\} \end{aligned} \tag{1.6}$$

then clearly $B \subset A$. In this particular example it is not true that $A \subset B$; that is, $A \not\subset B$. If for two sets $C \subset D$ and $D \subset C$ are both true, then $C = D$, that is, the two sets are equal (consist of the same elements).

Sets are conveniently illustrated schematically by *Venn diagrams*. A set is denoted by a plane area (the shape of the boundary is immaterial) bounded by a contour considered to contain the elements of the set. Figure 1 illustrates a variety of situations using Venn diagrams and allows us to introduce several other set concepts. The *union* (or *join* or *logical sum*) of two sets is written

$$A \cup B \qquad (\cup \text{ is called "cup"}) \tag{1.7}$$

and is itself a set consisting of all the elements of A *and* B. If we denote this set by C we can write

$$C = A \cup B \tag{1.8}$$

or

$$A \cup B = \{a \in C \mid a \in A \quad or \quad a \in B\} \tag{1.9}$$

where the vertical line is translated "with the property that." In Figure 1(b) all the elements inside *both* the circles denoted by A and B belong to the union of A and B. The *intersection* (or *meet* or *logical product*) of two sets is that set of elements common to both sets, and the corresponding

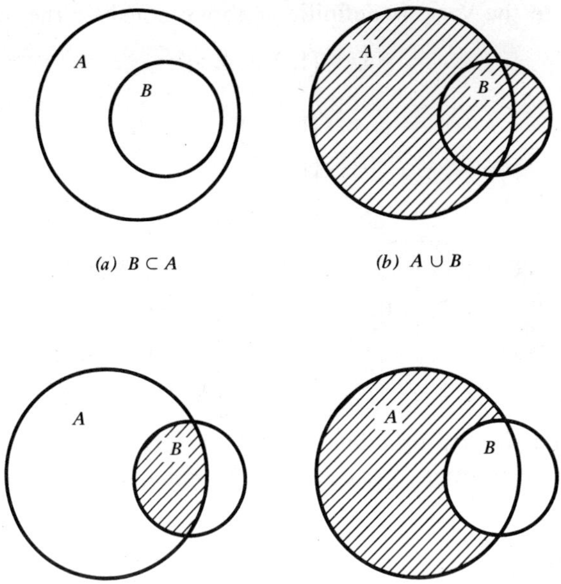

(a) $B \subset A$ (b) $A \cup B$ (c) $A \cap B$ (d) $A - B$

Figure 1

symbol is $\cap$ ("cap"). Thus, in Figure 1(*c*) the crosshatched area represents the intersection of A and B.

$$A \cap B = \{a \in C \mid a \in A \quad \text{and} \quad a \in B\} \tag{1.10}$$

Note the difference between the "or" and "and" in the definitions of union and intersection.

If two sets have no elements in common they are said to be *disjoint* and their intersection is the *null* or *empty* set $\varnothing$.

There are several other definitions which complete the set of manipulations we can perform with sets. As the union or logical sum combines elements of two sets (they may or may not be disjoint) we can denote the *difference* of two sets by $A - B$ by which we mean the set of elements belonging to A but not belonging to B. The *complement* of a set A is denoted by A' and is that set of elements not contained in A. One can *define* a *universal set* U to be that set containing all elements or all elements of particular interest; clearly elements outside a particular universal set cannot be considered in those manipulations involving that particular "universe." For any universal set U, its complement U' is clearly the null set $\varnothing$.

To illustrate the various definitions above consider the universal set

$$U = \{1, 2, 3, 4, 5\} \tag{1.11}$$

and the sets

$$\begin{aligned} A &= \{1, 2, 4\} \\ B &= \{1, 2\} \\ C &= \{3, 4\} \end{aligned} \tag{1.12}$$

Then

$$\begin{aligned} &B \subset A \\ &A \cup C = \{1, 2, 3, 4\} \qquad && B \cap C = \varnothing \\ &A \cap C = \{4\} && A' = \{3, 5\} \\ &A - C = \{1, 2\} && A \cap (B \cup C) = \{1, 2, 4\} = A \end{aligned} \tag{1.13}$$

An interesting *duality principle* exists for sets such that any identity involving the operations $\cap$ and $\cup$ remains valid if the $\cap$ and $\cup$ symbols are interchanged throughout.

Problems

1.1 Use Venn diagrams where necessary to verify the following laws of set algebra below where A, B, and C are subsets of the universal set U.

$$\begin{array}{ccc} U' = \varnothing & A \cap A' = \varnothing & \\ \varnothing' = U & A \cup A' = U & (A')' = A \\ A \cup A = A & A \cap A = A & \\ A \cup U = U & A \cup \varnothing = A & A - \varnothing = A \\ A \cap U = A & A \cap \varnothing = \varnothing & A - A = \varnothing \end{array}$$

$$\left.\begin{aligned} A \cap B = B \cap A \\ A \cup B = B \cup A \end{aligned}\right\} \quad \text{commutative laws}$$

$$\left.\begin{aligned} (A \cup B) \cup C = A \cup (B \cup C) \\ (A \cap B) \cap C = A \cap (B \cap C) \end{aligned}\right\} \quad \text{associative laws}$$

$$\left.\begin{aligned} A \cup (B \cap C) = (A \cup B) \cap (A \cup C) \\ A \cap (B \cup C) = (A \cap B) \cup (A \cap C) \end{aligned}\right\} \quad \text{distributive laws}$$

1.2 If a subset $B \subset A$ is such that at least one element of A is not in B then B is called a *proper* subset of A; the null set $\varnothing$ may also be considered a proper subset. The set A may be considered a subset of itself but is not a proper subset. Thus, the various subsets of $A = \{1, 2, 3\}$ are $\{1, 2, 3\}$, $\{1, 2\}$, $\{1, 3\}$, $\{2, 3\}$, $\{1\}$, $\{2\}$, $\{3\}$, and $\varnothing$. Use the binomal theorem to show that a set of n elements has 2^n subsets.

In our later discussion the idea of product groups will arise. This idea is based on the concept of the *product* of *sets* which we can define in several ways. First, consider an *ordered pair* of elements written as (a, b); if the set A consists of the n_A elements a_i ($i=1$ to n_A) and, similarly, $b_j \in B$ ($j=1$ to n_B), then the product of A and B can be written as $A \times B$ and consists of the $n_A n_B$ elements (a_i, b_j), that is, the $n_A n_B$ ordered pairs (a_i, b_j). Thus,

$$\begin{aligned} A &= \{a_1, a_2\} \qquad\qquad B = \{b_1, b_2, b_3\} \\ A \times B &= \{(a_1, b_1), (a_2, b_1), (a_1, b_2), (a_2, b_2), (a_1, b_3), (a_2, b_3)\} \end{aligned} \tag{1.14}$$

$B \times A$ in a like manner will consist of the same number of reversed ordered pairs (b_i, a_j). A simple example of the product of sets is the representation of a geometrical point in two dimensions (x_i, y_j). The set of all points in a particular rectangular area can be written as the product of sets representing the points on the abscissa and ordinate appropriate to the area in question. The product $A \times B$ is sometimes called the *Cartesian product.*

While not as common in set theory as the above Cartesian-type product, one can also define a product of two sets as the set of the products of the elements in the two sets. Clearly in such a case one must define what is meant by "multiplying" or taking the product of two elements in order to form such a product set. For example, if we were to specify ordinary multiplication as the product operation and were discussing sets of integers, then it would be clear that the product of the set $\{1, 2\}$ and $\{1, 5, 10\}$ is the set $\{1, 2, 5, 10, 20\}$ consisting of all the *distinct* products. We will be much concerned with this type of product for sets having the group properties. That it is much less general in set theoretical considerations than the Cartesian product is evident in that a combining principle must be specified. The presence of such a combining principle is an example of set structure which we now take up.

A set may have an arbitrary membership or may be characterized by some feature common to or relating the elements; examples are $A = \{a;\, a$ is a positive integer$\}$ or $A = \{a;\, a$ is a positive power of $2\}$ or $A = \{a;\, a$ multiplied by itself is unity$\} = \{1, -1\}$. If there is a system of relationships between the elements of a set the set is said to possess *structure,* the structure being the particular set of relationships. A set may possess *order* structure such that $a < b < c < \cdots$; the elements of the set $\{1, 2, 3, 5\}$ clearly can be ordered in ascending magnitude $1 < 2 < 3 < 5$ and therefore the set has order structure. In the study of groups we are most concerned with sets possessing *algebraic* structure, that is, sets where there is a *composition law* for any two elements of the set. The composition law is generally referred to as *multiplication;* this does not necessarily mean ordinary multiplication but can be addition, subtraction, ordinary multiplication, or any other defined procedure for multiplying or combining any two elements of the set in question. The set $\{0, 1, 2, 3\}$ where the composition law is "addition modulo 4" has algebraic structure, as does the set $\{1, -1, i, -i\}$, where $i = (-1)^{1/2}$, and the composition law is ordinary multiplication. Note that, in both examples, combining any two elements of the sets by the appropriate rule of multiplication leads to another element in the set. Too, an element may be combined or multiplied with itself and the resulting product must also reside in the set if the set has algebraic structure. We denote the composition of a, $b \in A$ by

$$a \cdot b \qquad \text{or} \qquad ab = c \in A \tag{1.15}$$

where in the first instance the dot signifies the multiplication rule. It is often omitted and we shall do so using the second alternative above. The order of the elements in a product can be significant and generally

$$ab \neq ba \tag{1.16}$$

that is, the elements a and b may not *commute*. If the order *is* immaterial we say that a and b commute. Whether or not two elements commute may depend upon the particular elements involved as well as upon the composition law. Consider the set of all integers under addition (i.e., addition is the composition law); then clearly

$$(1) + (-3) = (-3) + (1) = (-2) \tag{1.17}$$

and it is easy to see that all elements of the set commute. However, if the composition law were subtraction then

$$(1) - (-3) = (4) \neq (-3) - (1) = (-4) \tag{1.18}$$

and in this instance two elements generally do not commute. It *is* true in the latter case that for $a \in A$, $(a) - (a) = (0) \in A$, and each element commutes with itself. This is a rather obvious case but it illustrates the fact that a set with algebraic structure can have mixed commutation rules depending on the elements involved. Generally speaking, regardless of the composition law any element will commute with itself.

A better example is provided by using as elements of a set those symmetry operations that characterize the symmetry of some geometrical object. Here the composition law is operator multiplication; we are discussing a set of *operators* where equalities, such as $ab = c$, imply that the operation (on the geometrical object) by b followed by the operation by a yields the same result as the operation by c. Consider the point group C_{4v} consisting of the set of eight elements

$$C_{4v} = \{E, C_4, C_2, C_4^3 = \bar{C}_4, \sigma_v, \sigma_v', \sigma_d, \sigma_d'\} \tag{1.19}$$

Then

$$C_2C_4 = C_4C_2 = C_4^3 \tag{1.20}$$

and the elements C_2 and C_4 commute. On the other hand,

$$\sigma_v C_4 = \sigma'_d \neq C_4\sigma_v = \sigma_d \tag{1.21}$$

and σ_v and C_4 do not commute.

Another property of combining elements in a set with algebraic structure is whether or not the associative law holds. Thus $a(bc)$ is interpreted as performing the set multiplication rule on bc followed by multiplication by a; whereas, $(ab)c$ represents the operation c multiplied or combined with the result of the operation ab. For ordinary multiplication or addition, the associative law holds

$$a(bc) = (ab)c \tag{1.22}$$

as in the case of addition

$$a + (b + c) = a + b + c = (a + b) + c = (ab)c \tag{1.23}$$

Conversely, for subtraction or division the associative law does not obtain

$$\begin{aligned} \text{subtraction:} \quad & a(bc) = a - (b - c) = a - b + c \\ & (ab)c = (a - b) - c = a - b - c \\ \text{division:} \quad & a(bc) = a \div (b \div c) = ac/b \\ & (ab)c = (a \div b) \div c = a/bc \end{aligned} \tag{1.24}$$

Whereas group elements may or may not commute, one defining feature of a group is that the allowable composition laws must be such that the associative law always holds. Another way of stating the associative law is to say that the order of the elements in a product uniquely determines the result; the manner in which the elements are grouped (by the brackets) is immaterial if the associative law obtains.

A final feature of sets that will be discussed later is that of *mapping*. When there is a rule or law that connects an element $a \in A$ to a *uniquely determined* element $b \in B$, we say a mapping exists between the sets A and B. Knowledge of the basic ideas of mapping is useful in relating abstract and real groups, particularly those real groups we call *representations*. This topic will be taken up after our preliminary discussion of groups.

SOLUTIONS TO PROBLEMS

1.1 Answers given in problem.

1.2 The total number N of subsets including the set itself (and the null set) is given by counting the number of ways of selecting r objects from n objects and summing r from 0 to n.

$$N = \sum_{r=0}^{n} \binom{n}{r}$$

Where $\binom{n}{r} = n!/(n-r)!\,r!$ is the binomial coefficient. The binomial theorem is

$$(x+y)^n = \sum_{r=0}^{n} \binom{n}{r} x^r y^{n-r}$$

Setting $x = y = 1$ yields

$$2^n = \sum_{r=0}^{n} \binom{n}{r} = N$$

2
Groups

DEFINITION OF A GROUP

A group is a set with algebraic structure whose composition law satisfies four basic postulates:

1. A multiplication (or composition) law exists and the group is *closed* under the group operation. The last part of this statement implies that the product of any two elements in the group (including multiplication of an element by itself) is also an element of the group. If the group is signified by G, then for $a, b \in G$, $ab \in G$, all a, b.
2. The associative law holds.

$$a(bc) = (ab)c \tag{2.1}$$

   The order of the elements in a product uniquely determines the result.
3. There exists a unique identity (or unit) element, $e \in G$, such that for all elements a of the group (including e itself)

$$ae = ea = a \tag{2.2}$$

   (Capital "E" is usually used to denote the identity symmetry operation; we will use lower case "e" to denote the identity element in an abstract group.)
4. Each element a has a unique inverse a^{-1}, also contained in the group such that

$$a^{-1}a = aa^{-1} = e \tag{2.3}$$

   The inverse of a^{-1} is clearly a.

Note that we have not presumed that the commutative law holds. If a group is such that this law *does* hold—that is, that all elements of the group commute—the group is called *abelian* (after N. H. Abel). An example of an abelian group would be the symmetry point group C_4. (Sets that obey postulate 1 above are called grouppoids; if both postulates 1 and 2 are obeyed, the set is called a semigroup.)

The number of elements in the group is called the *order* of the group and will be denoted $^\circ G$. Just as with sets one may have finite or infinite groups; whereas the point group C_4 is a finite group, the point group C_∞ is obviously an infinite group.

Problems

2.1 Prove that the inverse of an element is unique.

2.2 Show that the inverse of a product is given by the product of the separate inverses in reverse order

$$(abc)^{-1} = c^{-1}b^{-1}a^{-1}$$

MULTIPLICATION TABLES

For a group of ^{o}G elements there will be $(^{o}G)^2$ products, clearly not all different since then there would be more than ^{o}G distinct members (elements) of the set. We can completely specify the structure of a group by displaying the $(^{o}G)^2$ products in what is called the group multiplication table or simply the group table. We do so as illustrated below where the various ^{o}G elements label both the columns and rows; the product of the two labels is at the intersection of a given column and row.

G	e	a	b	c	$\cdots$
e	e	a	b	c	
a	a	a^2	ab	ac	
b	b	ba	b^2	bc	
c	c	ca	cb	c^2	
$\cdot$					
$\cdot$					
$\cdot$					

We adopt the convention as illustrated in the table of taking the product to be (row label) · (column label). The first row and column reflect the property of e, the identity element, which, again by our convention, will always be the initial label of both the rows and columns. Note that an abelian group will reveal itself by being symmetrical across the group table diagonal.

A particularly useful property of the group table is that a given element occurs once and only once in any given row or column. This means that the rows and columns are simultaneously rearrangements of some one basic row and column and is sometimes referred to as the *rearrangement theorem*. The proof of this statement is straightforward and follows closely that used to prove the uniqueness of the identity element. One proceeds by assuming the contrary and then exhibiting a contradiction.

Take, for example, the second row in the above table and assume a given element occurs twice, that is, two products in that row are identical. But, if, for example,

$$\begin{aligned} ab &= ac \\ a^{-1}ab &= a^{-1}ac \\ b &= c \end{aligned} \tag{2.4}$$

b and c, supposedly distinct elements, are equal. This contradicts one of the basic presumptions of sets and groups; therefore, the assumption is incorrect and the desired theorem is proved.

As an example of a group, consider the sets $\{i, -1, -i, 1\}$ and $\{1, -1\}$ under ordinary multiplication. It is easily verified that the group properties are satisfied and that the multiplication tables are given by (1 is the identity element in both cases)

	1	i	-1	$-i$
1	1	i	-1	$-i$
i	i	-1	$-i$	1
-1	-1	$-i$	1	i
$-i$	$-i$	1	i	-1

	1	-1
1	1	-1
-1	-1	1

We see that each element appears only once in a given row or column and further that these particular examples are both abelian groups. The first is of order 4 while the second is of order 2. In addition these two groups are also examples of what are called *cyclic* groups. A cyclic group is one whose elements are all given by one particular element (the *generating element*) and all its distinct powers. In the first instance the group $\{1, i, -1, -i\}$ is generated by i (or $-i$)

$$i, i^2 = -1, \qquad i^3 = -i, \qquad i^4 = 1 \tag{2.5}$$

while in the second case, -1 is the generating element

$$-1, (-1)^2 = 1 \tag{2.6}$$

Note that higher powers do not yield any *new* elements and that the identity element is given by the generating element raised to a power equal to the order of the cyclic group. This will generally be so for a cyclic group

$$G = \{a, a^2, a^3, \ldots, a^n = e\} \qquad {}^{o}G = n \tag{2.7}$$

It is easy to see that one particular way of producing a cyclic group of order n under ordinary multiplication is to choose as the generator the nth root of unity, $e^{2\pi i/n}$.

Any cyclic group is specified by its order; thus, aside from using different symbols, all cyclic groups of order 4 must basically be the "same," as the above example, and must have the same structure. Two groups which have the same structure, regardless of whether they are cyclic or not, are said to be *isomorphic*.

Problems

2.3 Construct abstract groups of $^{\circ}G=2$ and $^{\circ}G=3$ and show that they are unique.

2.4 Consider the group of integers where the composition law is addition modulo 4 (mod 4). Show that this group is isomorphic to the cyclic group generated by *i*. Find the correspondence between these two groups.

2.5 Work out the multiplication table for the point group C_{3v}.

2.6 Consider the set of all integers under addition. Do they form a group? Would a group exist with the element 0 and positive integers only?

2.7 Consider the set of all integers under subtraction. Do they form a group?

SUBGROUPS

In our previous discussion of sets the idea of subsets was introduced. If $B \subset A$ then B is a subset of A. If A and B are both groups under the same law of composition and if $B \subset A$ then B is said to be a *subgroup* of A. Let $B \subset A$ where A possesses algebraic structure. B is called a *stable* subset of A if $bb' \in B$ for any $b, b' \in B \subset A$ under the law of composition of A. The subset B is thus *closed* under multiplication. If A is a group and B is a stable subset of A, then B will be a subgroup of A if the elements of B satisfy the group postulates.

We can define a subgroup in an alternative and concise fashion: A nonempty subset H of a group G is a subgroup of G, if and only if with each pair of elements $h, h' \in H$, also $h^{-1}h' \in H$. The associative law holds since $H \subset G$. The identity is present since for $h' = h$, $h^{-1}h = e \in H$. Inverses are present since for $h' = e$, $h^{-1}e = h^{-1} \in H$. Finally, the subset is stable since for $h = h^{-1}$, $(h^{-1})^{-1}h' = hh' \in H$. Note that one can define a *group* in this way by considering every group as a subgroup of some larger group.

Most of our considerations will involve finite groups. For finite subsets of a group *only stability* is necessary for the subset to be a group (subgroup); this is a most useful criterion to apply to classify such subsets as subgroups. Such is not the case for infinite groups. For example, consider the group of all integers under addition. The subset consisting of all positive integers plus zero is a stable subset of the group but is not a subgroup since inverses are not present. Nor is stability of a set enough; thus, subtraction to some modulus constitutes a stable set, but the associate law is not generally obeyed.

G may be considered a subgroup of itself but is called a *trivial* or *improper* subgroup. The group $\{e\}$ consisting of simply the identity element is also a *trivial* subgroup of any group. All other subgroups are called *proper*.

Problems

2.8 Use the finiteness of the set to show that any stable subset of a finite group is a subgroup.

2.9 Find the subgroups of the cyclic group of order 4.

2.10 Find the subgroups of the point group C_{3v}.

2.11 The *order* n of an element h of a group is that power of h such that $h^n = e$. Show that for any element $h \in G$ one can form a cyclic subgroup of order n by collecting h and all its distinct powers.

SOLUTIONS TO PROBLEMS

2.1 Prove by contradiction. Assume

$$ab = e \qquad \text{and} \qquad ac = e \qquad \text{where} \qquad b \neq c$$

But then

$$ab = ac$$
$$a^{-1}ab = a^{-1}ac$$
$$b = c$$

which is a contradiction since all elements of a group (or set) are presumed different.

2.2 Let $d = abc$. Then

$$dc^{-1} = abcc^{-1} = ab$$
$$dc^{-1}b^{-1} = a$$
$$dc^{-1}b^{-1}a^{-1} = aa^{-1} = e$$

and clearly

$$d^{-1} = (c^{-1}b^{-1}a^{-1}) = (abc)^{-1}$$

Note that it is important to multiply both the left- and right-hand sides of the above expressions on the same "side." In this example we right-multiplied; the problem could also be solved by left-multiplying in the sequence a^{-1}, b^{-1}, and then c^{-1}.

2.3 For $°G=2$ let $G=\{e, a\}$. The group table is given below.

	e	a
e	e	a
a	a	e

where $a^2=e$ must obtain according to the theorem regarding the appearance of elements in rows and columns. For $°G=3$ let $G=\{e, a, b\}$ so partially we have

	e	a	b
e	e	a	b
a	a		
b	b		

In the second row the next two entries must be e, b or b, e; however, the first case puts two b's in the last column and must be discarded. Or, stated alternatively, if $ab=b$ this implies $a=e$ which cannot be so. Filling out the remainder of the table follows easily to give

	e	a	b
e	e	a	b
a	a	b	e
b	b	e	a

In both examples there is only one solution allowed, so that the abstract groups of orders 2 and 3 are unique. The abstract order 1 group, simple $\{e\}$, is obviously unique; for $°G=4$ we find two different abstract groups, and higher-order groups have increasingly more solutions for a given order. Note that the above orders 2 and 3 are both cyclic groups; one can show that groups of prime order must be cyclic.

2.4 The two multiplication tables are

addition mod 4

	0	1	2	3
0	0	1	2	3
1	1	2	3	0
2	2	3	0	1
3	3	0	1	2

cyclic $°G=4$

	1	i	-1	$-i$
1	1	i	-1	$-i$
i	i	-1	$-i$	1
-1	-1	$-i$	1	i
$-i$	$-i$	1	i	-1

and the correspondence is that given by the order of labeling

$$0 \leftrightarrow 1 \qquad \text{(identity elements)}$$
$$1 \leftrightarrow i$$
$$2 \leftrightarrow -1$$
$$3 \leftrightarrow -i$$

In the addition mod 4 group, the identity element is 0 and the various inverses are

$$(0)^{-1}=0 \qquad (1)^{-1}=1$$
$$(1)^{-1}=3 \qquad (i)^{-1}=-i$$
$$(2)^{-1}=2 \qquad (-1)^{-1}=-1$$
$$(3)^{-1}=1 \qquad (-i)^{-1}=i$$

2.5 With the stereographic projection as shown below (or by the triangle displaying the pertinent symmetry elements) and taking positive rotations to be in a clockwise direction, we obtain the group table (Table 1) as given (see also Figure 1).

Table 1

C_{3v}	E	σ_3	σ_2	σ_1	C_3	C_3^2
E	E	σ_3	σ_2	σ_1	C_3	C_3^2
σ_3	σ_3	E	C_3^2	C_3	σ_1	σ_2
σ_2	σ_2	C_3	E	C_3^2	σ_3	σ_1
σ_1	σ_1	C_3^2	C_3	E	σ_2	σ_3
C_3	C_3	σ_2	σ_1	σ_3	C_3^2	E
C_3^2	C_3^2	σ_1	σ_3	σ_2	E	C_3

This group of order 6 will be used in further problems examples and is accordingly important to understand.

2.6 Yes. Associative law holds, the identity element is 0

$$(n)+(0)=(n)$$

and the inverse of each element is its negative

$$(n)+(-n)=(0)$$

We could not have a group if zero were not present, because we would lack an identity. Furthermore, if all the negative integers were removed, elements would have no inverses and a group would not exist.

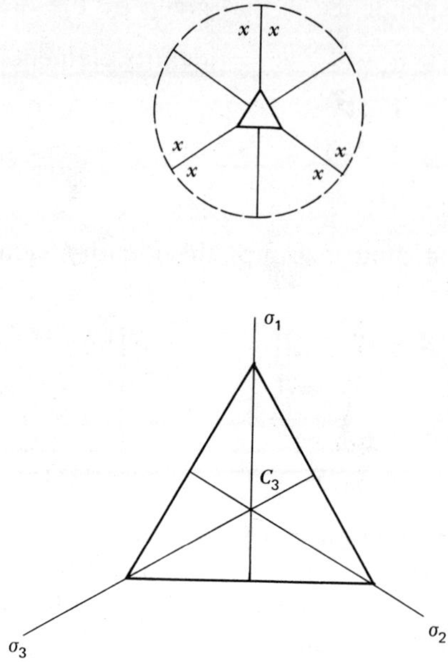

Figure 1

2.7 A *right identity is present* (*the element zero*)

$$(n)-(0)=(n)$$

but it is not a *left* identity

$$(0)-(n)=(-n)\neq(n)$$

Indeed, a left identity does not exist. Both must be present and equal to each other for a group to exist. Furthermore, the associative law does not obtain. Although this set possesses algebraic structure it is not a group.

2.8 For $h \in H \subset G$, if G is finite, so too must H be finite. Therefore, there must exist some power of h (the *order* of the element h) such that $h^n = e$, since otherwise the set would not be stable. If $h^n = e$, then $h^n = h^{+1}h^{n-1} = e$ and it is clear that $h^{-1} = h^{n-1}$ is present and the characteristics of a group are present.

2.9 Let $G = \{a, a^2, a^3, a^4 = e\}$. The subgroups are

$\{e\}$, G, the trivial subgroups, and

$\{e, a^2\}$, a proper subgroup

2.10 Aside from the trivial subgroups one finds the proper (nontrivial) subgroups

$$\{E, \sigma_1\}, \{E, \sigma_2\}, \{E, \sigma_3\}, \{E, C_3, C_3^2\}$$

2.11 If the order of $h \in G$ is n, then the set of all distinct powers of h is $\{h, h^2, h^3, \ldots, h^n = e\}$ and is clearly a subgroup of order n. Note that the four subgroups of C_{3v} (problem 2.10) are of this type.

3
The Partitioning and Factoring of Groups

COSETS AND LAGRANGE'S THEOREM

Cosets of subgroups allow us to divide a group in a particularly useful way into disjoint sets. We need the concept of cosets in order to discuss *factor groups* later on and presently to prove *Lagrange's theorem.* Lagrange's theorem states that

$$\text{if } H \text{ is a subgroup of } G \text{ of order } {}^{\circ}H, \text{ then}$$
$${}^{\circ}G \div {}^{\circ}H = \text{an integer}$$

that is, the order of G must be an integral multiple of the order of H for *all* $H \subset G$.

Lagrange's theorem is obviously true for the trivial subgroups $\{e\}$ and G. Consider a nontrivial or proper subgroup H; since H is proper, there must exist some element a belonging to G but not belonging to H. Multiply all the elements of H on the left by a and collect all those products into a set symbolized by aH.

$$aH = \{ae, ah_1, ah_2, \ldots\} = \{ah_1 \mid h_i \in H \quad \text{and} \quad a \notin H\} \tag{3.1}$$

aH is called the *left coset of H by a.* In a similar manner we can define the right coset of H by a

$$Ha = \{h_i a \mid h_i \in H \quad \text{and} \quad a \notin H\} \tag{3.2}$$

In general, left and right cosets of H by a will not be the same.

Cosets have a number of important properties. We will consider the left coset of H by a to illustrate these although it should be clear that the same properties exist for right cosets.

1. aH is not a group because e is not contained. If e were contained, then for some h_i, $ah_i = e \Rightarrow a = h_i^{-1} \in H$ which violates our condition that $a \notin H$. Here the symbol "$\Rightarrow$" means "implies" or "leads to."
2. $(aH) \cap (Ha) \neq \varnothing$. Both aH and Ha contain the element a.
3. $(aH) \cap H = \varnothing$. Otherwise, $aH_i = h_j \quad a = h_j h_i^{-1} \in H$. Thus, all the elements of aH are different from those of H.

4. All the elements of aH are distinct, since, otherwise, $ah_i = ah_j \Rightarrow h_i = h_j$. Thus aH contains $^\circ H$ distinct elements, the same number as in H itself.
5. Two left cosets of a given subgroup are either disjoint or identical. If aH and bH are two left cosets and if $ah_i = bh_j$, then $a = bh_jh_i^{-1} = bh_k \in bH$ and $a \in bH$. Thus, for *any* element $ah_k \in aH$, $ah_k = bh_jh_i^{-1}h_k = bh_m \in bH$.

Property 5 shows us that all the distinct left cosets of H along with H itself divide or partition the group G into a set of disjoint sets. This division of G into disjoint sets is appropriately called the *left quotient set of G by H.*

Since each coset aH contains $^\circ H$ elements and the quotient set *spans* the group G (i.e., contains all the elements of G) then the number of sets in the quotient set multiplied by $^\circ H$ is equal to $^\circ G$. The numbers involved above are obviously integers and Lagrange's theorem is proved. The number of sets in the quotient set—the number of distinct cosets aH plus H itself—is called the *index of H in G.* It should be clear that the index of H in G does not depend on whether we consider left or right quotient sets. The spanning of G by H and its distinct cosets (left or right) can be written

$$(aH)\cup(bH)\cup(cH)\cdots = G = (Ha)\cup(Hb)\cup(Hc)\cdots \tag{3.3}$$

As an example, consider the noncyclic group of $^\circ G = 4$. This is one of the two distinct groups of order 4 (the other is the cyclic group of order 4); it is often symbolized by V and is often simply called the four group.

	e	a	b	c
e	e	a	b	c
a	a	e	c	b
b	b	c	e	a
c	c	b	a	e

The proper subgroups and their cosets are given below.

$$\begin{aligned} H_1 &= \{e, a\} & bH_1 &= \{b, c\} \\ & & cH_1 &= \{c, b\} = bH_1 \\ H_2 &= \{e, b\} & aH_2 &= \{a, c\} = cH_2 \\ H_3 &= \{e, c\} & aH_3 &= \{a, b\} = bH_3 \end{aligned} \tag{3.4}$$

In the example above the reader can verify that the left and right cosets for a particular subgroup are the same. This is generally not so. If it is, that is, if

$$sH = Hs \qquad \text{all} \qquad s \in G \tag{3.5}$$

H is said to be a *normal* subgroup.

Problems

3.1 Recalling exercise 2.11 show that the order of any group element of a finite group is a divisor of the order of G.

3.2 Show that any group of prime order is necessarily cyclic.

3.3 Show that any subgroup of index 2 is normal.

3.4 Show that any subgroup of an abelian group is normal.

3.5 Find the left and right cosets of $H_1 = \{e, \sigma_1\}$ and $H_2 = \{e, C_3, C_3^2\}$ in C_{3v}.

Lagrange's theorem, combined with the fact that an element's order must be a divisor of the group (see problem 3.1 above), is very useful in determining the possible structure of a group. Thus a group of order 4 can only have proper subgroups of order 2; its elements may only have order 1, 2, or 4. A group of order 6 can only have nontrivial subgroups of order 2 or 3; its elements must have order 1, 2, 3, or 6. Consider the possible structures for groups of order 4. If an element of order 4 exists the group is necessarily cyclic and its multiplication table is

	e	a	a^2	a^3
e	e	a	a^2	a^3
a	a	a^2	a^3	e
a^2	a^2	a^3	e	a
a^3	a^3	e	a	a^2

or

	e	a	b	c
e	e	a	b	c
a	a	b	c	e
b	b	c	e	a
c	c	e	a	b

where in the second group table we have rewritten $a^2 = b$ and $ab = a^3 = c$, simply another way of representing the abstract cyclic group of order 4. Consider the case where there is no element of order 4; then each element must have order 2 except e which is always of order 1. Thus, for this second case

$$G = \{e, a, b, c\}; \qquad a^2 = b^2 = c^2 = e \tag{3.6}$$

and the group table will be

	e	a	b	c
e	e	a	b	c
a	a	e	c	b
b	b	c	e	a
c	c	b	a	e

where we have made use of the rearrangement theorem.

Problems

3.6 Determine the structure of a noncyclic group $°G=6$ where an element of order 3 is present. Employ the fact that all elements must be of order 2 or 3.

3.7 For a normal subgroup $Hs=sH$, all $s \in G$. This implies $s^{-1}Hs=H$. Show that the elements of $s^{-1}Hs$ do indeed form a group.

FACTOR GROUPS

The left and right quotient sets of a normal (or *invariant*) subgroup are the same. In addition, these sets of elements treated in a particular way display the properties of a group. Let N be a normal subgroup and denote the quotient set as

$$N, aN, bN, \ldots = N, Na, Nb, \ldots \tag{3.7}$$

The product of an element in aN and an element in bN is an element in the set abN; since N is normal

$$b^{-1}n_ib = n_k \in N \tag{3.8}$$

so that

$$(an_i)(bn_j) = abb^{-1}n_ibn_j = abn_kn_j \in abN \tag{3.9}$$

The above is true for any two elements of aN and bN and is symbolized

$$(aN)(bN) = abN \tag{3.10}$$

If we consider the cosets of an invariant subgroup as entities in themselves, the product of two sets of elements in the quotient set is also a set of elements in the quotient set and the quotient set forms a group. Thus

$$\begin{aligned} (N)(N) &= N \\ (aN)(N) &= aN \\ (aN)(a^{-1}N) &= N \end{aligned} \tag{3.11}$$

Inverses are present and the normal subgroup N plays the role of the identity or unit element. This quotient set G/N is called the *factor group* of G by N. Its order is clearly $°G/°N$ or the index of N in G.

Consider the four group $V=\{e, a, b, c\}$ and the normal subgroup $N=\{e, a\}$

	e	a	b	c
e	e	a	b	c
a	a	e	c	b
b	b	c	e	a
c	c	b	a	e

$$bN = \{be, ba\} = \{b, c\}$$
$$V = N \cup bN$$

N and bN (or cN) form the factor group, and we can easily write its multiplication table

V/N	N	bN
N	N	bN
bN	bN	N

where $bNbN = N$ because $b^2 = e$ and $bn_ibn_j = b^2n_kn_j = n_kn_j \in N$. Note that in this example both $\{e, b\}$ and $\{e, c\}$ are also normal subgroups since they are of index 2. We could have formed factor groups with these normal subgroups but the same factor group structure would result.

$\{e\}$ is a (trivial) subgroup of G and is obviously normal. The factor group of G by $\{e\}$ is simply G itself. In a like manner the factor group G/G is the order 1 group $\{e\}$.

We see that the various factor groups of a group are defined by the structure of the group; conversely, knowledge of the factor groups helps in elucidating the structure of the parent group. In forming factor groups (or, generally, quotient sets) certain associations of sets of elements have been made, which, as we see later, define a mapping of one group to another group. A special relationship exists between allowed mappings of this kind and factor groups which we shall discuss later.

Problem

3.8 Find the factor groups of the point group C_{3v}.

PRODUCTS OF GROUPS

We have seen how sets and groups can be partitioned or divided into subsets. In the case of groups we have thus far discussed subgroups and cosets (remember that a coset is *not* a group); aside from the group G itself, subgroups of G must have an order lower than $^{\circ}G$. In a similar way, one can define the product of groups which, as will become clear, have orders equal to or greater than the orders of the groups involved in the product.

There are two basic types of product groups depending upon whether or not there is an external law of composition, that is, a definition of the product of an element in one group with an element in the other.

We will begin with the situation where there exists a single composition law for the groups under consideration. Consider two groups A and B. The *product set* $C = AB$ consists of the elements $c = ab$ with $a \in A$, $b \in B$, where all possible products ab of elements are considered. The set C is not necessarily a group; for if $a_ib_i, a_jb_j \in C$, then $(a_ib_i)^{-1}(a_jb_j) = b_i^{-1}a_i^{-1}a_jb_j$ is not generally of the proper form ab, an element in A

multiplying an element in B. The group condition *will* be satisfied if A and B commute, $AB = BA$ (that is, for each element $a_i b_j$ in AB, the same element exists in BA as some $b_k a_k$), for then

$$\begin{aligned}(a_i b_i)^{-1}(a_j b_j) &= (b_k a_k)^{-1}(a_j b_j)\\ &= a_k^{-1}(b_k^{-1} a_j) b_j\\ &= a_k^{-1} a_q b_q^{-1} b_j \in AB\end{aligned} \tag{3.12}$$

the commuting property of A and B being used twice above. Thus, if the groups A and B commute, their *product*, AB, will be a group. It should be clear that since both A and B contain e, and since AB contains both $a_i e$ and eb_j (all $a_i \in A$ and $b_j \in B$), A and B separately will be subgroups of $C = AB$.

Generally, one will not know readily if A and B commute. We can see, however, that for $A, B \subset C$ if A or B is a normal subgroup of the group G, A and B will commute and the product will be a subgroup of G; if both A and B are normal, the product AB will also be a normal subgroup of G.

The more important type of product of groups with a common composition law is the *direct product* which is written as $A \times B$. A group C is a direct product of its subgroups A and B, $C = A \times B$, if

1. $C = AB$ (simple product relationship),
2. for each element $c \in C$ the elements $a \in A$ and $b \in B$ of the product $c = ab$ are uniquely determined,
3. $ab = ba$ for all $a \in A$, $b \in B$.

Note the difference between product and direct product in terms of the commutation rules that must be satisfied. In the case of the *product*, the *subgroups* had to commute in the sense that $AB = BA$, while in the case of the *direct product* all products of elements ab must commute. If C is a direct product of its subgroups A and B, both A and B are said to be *direct factors* of C. It follows from the direct-product requirements that A and B are normal subgroups and that their intersection contains only the element e; these two statements can, in fact, replace (2) and (3) above and are somewhat easier to state (see problem 3.9).

As an example of a direct product, consider the cyclic group of order 6 ($a^6 = e$). It can be written as the direct product of A and B where

$$\begin{aligned}A &= \{e, a^2 a^4\}\\ B &= \{e, a^3\}\end{aligned} \tag{3.13}$$

Clearly $C = AB$ and all the elements of A and B commute; further, all $c = ab \in C$ can be written as ab products in only one way. Expressing the above result in terms of points groups, we have shown that $C_6 = C_3 \times C_2$.

We note finally that for the product or direct-product group, its order is equal to the product of the orders of the groups in the product or direct product. It should be clear also that multiple products such as $ABC\cdots$ and $A\times B\times C\times\cdots$ can readily be defined by simply extending the processes described above. Since the various products are commutative the order of the factors is immaterial,

$$AB = BA, \qquad A\times B = B\times A$$

Problems

3.9 Show that the defining conditions for the direct product $G = H\times K$ imply that H and K are normal and $H\cap K = \{e\}$.

3.10 Determine if the point group C_{2h} can be written as a direct product of two of its subgroups.

3.11 Determine if the point group C_{3v} can be written as a direct product.

3.12 What point groups are formed from the direct products
a. $C_3\times C_i$ and
b. $C_3\times C_s$?

The other type of product of groups is called the *abstract direct product* and is used without requiring a common multiplication law for the two groups involved. Let A and B be two arbitrary groups (with arbitrary composition laws); then $C = A\times B$ (unfortunately, the same notation as in the direct product) will be a group where the elements of C are the *ordered pairs* of elements (a_i, b_j), $a_i\in A$, $b_j\in B$. Multiplication is carried out between elements belonging to the same initial group:

$$(a_1, b_1)(a_2, b_2) = (a_1a_2, b_1b_2) \tag{3.14}$$

The identity and inverses are readily seen to be

$$\begin{gathered}(e(\text{from } A), e(\text{from } B))\\ (a_1, b_1)^{-1} = (a_1^{-1}, b_1^{-1})\end{gathered} \tag{3.15}$$

The order of C is, again, the product of the orders of A and B: ${}^{\circ}C = {}^{\circ}A{}^{\circ}B$. As with the product and direct-product groups, the abstract direct-product group can be extended to more than ordered pairs in an obvious way; thus, the abstract direct product of n groups could consist of ordered $n$mers and its order would be ${}^{\circ}A{}^{\circ}B{}^{\circ}C\cdots{}^{\circ}N$.

Note that in this type of abstract direct product, the set of elements (e, b_i), all $b_i\in B$ forms a subgroup S_B of C; the same is true for the set (a_i, e), all $a_i\in A$, yielding the subgroups S_A of C. S_A and S_B are both normal, and one can show that both are direct factors of C; that is, $C = S_A\times S_B$ is the (ordinary) direct product of S_A and S_B.

The abstract direct product is similar to the Cartesian product of sets defined earlier where one spoke of ordered pairs of elements. Its use arises naturally in systems where a mixture of symmetries is present.

The groups that are important in our physical situations are those groups of operators that commute with the system's Hamiltonian. Consider the physical example of a water molecule. In this system the coordinates of the hydrogen nuclei can be interchanged, the coordinate of the electrons can be interchanged, and the molecule can be rotated, all without changing the energy of the system. These various sets of operations will each form a group, and the elements of each group clearly commute with each other. Accordingly, an abstract direct-product group may be formed where the elements of the group would be the ordered trimers of the elements from the groups of nuclear and electron permutations and the group of molecular rotations, in some specific but arbitrarily selected order.

The abstract direct product is really the simplest type of product to understand. For if a system can be characterized by two groups of operations, A and B, referring to different parts of the total system, then the ordered pairs in $A \times B$, (a_i, b_j) are simply the set of combined operations "do a_i to subsystem A and b_j to subsystem B." This way of expressing the element (a_i, b_j) should make it clear how the ordered pair or, in general, the ordered $n$mer can be considered a single element of the larger product group.

CONJUGATE ELEMENTS AND CLASSES

Although the term *class* in set theory has a broader definition than that given below, it is fairly universal among scientists to restrict its use to what are called *conjugation classes*, and the latter is what we shall mean when speaking of a class in a group. A group may be partitioned into classes just as a group may be partitioned into left or right quotient sets. When we discuss *representations* of groups the class concept will be most useful, and it is primarily for this reason that the idea is introduced here. In addition, when speaking of groups of operators (such as symmetry operations of a point group) the group elements (operators) in the same class have a conceptually useful relationship to each other in terms of coordinate transformations.

Consider a group G and two elements $a, b \in G$. If there exists an element $s \in G$ such that

$$a = s^{-1}bs \tag{3.16}$$

a is said to be *conjugate to b by the element s*. Clearly $b = sas^{-1}$ and b is conjugate to a by s^{-1}. Combining those two statements, we simply say that a and b are conjugate. If we select some element $g \in G$ and form all the (distinct) products $s^{-1}gs$ for all $s \in G$ we generate a class of elements, all of which are conjugate to one another. Thus, if a and b are elements of the same class (the class of the element g)

$$\begin{aligned} a &= s^{-1}gs \\ b &= k^{-1}gk \end{aligned} \tag{3.17}$$

and

$$a = s^{-1}gs = s^{-1}kbk^{-1}s = (k^{-1}s)^{-1}b(k^{-1}s) \tag{3.18}$$

It is easily shown (see problem 3.13 below) that if a is conjugate to b and b is conjugate to c that also a is conjugate to c; this is known as the *transitive property* of classes. This transitive property shows us that the classes of a group are disjoint (no common members) and that a class, denoted C_g or $[g]$, may be specified by labeling it with any element in the class

$$[g] \equiv C_g = \{g_i \in G \mid g_i = s^{-1}gs, \quad \text{some} \quad s \in G\} \tag{3.19}$$

Classes have a number of identifying features, some of which are given in the exericses below. One property, which we state without proof, is that the order of a class ${}^{o}C_g$ (the number of elements in the class) is a divisor of ${}^{o}G$ the order of the group (see Appendix A). Thus, the orders of classes in a group ${}^{o}G = 6$ may only be 1, 2, 3, or 6. Actually, for ${}^{o}G > 1$ groups one will not have classes ${}^{o}C_g = {}^{o}G$ as can be deduced from problem 3.14 below.

Problems

3.13 Show that if $a \sim b$ (a is conjugate to b) and if $b \sim c$ that $a \sim c$.

3.14 Show that e the identity element is always in a class by itself.

3.15 Show that the orders of all elements in a given class are the same.

To illustrate the formation of classes consider the point group C_{3v} and its multiplication table as given in the solution to problem 2.5. E is always in a class by itself: $[E]$. Consider the class $[\sigma_1]$. Forming all the appropriate products,

$$\begin{aligned} E^{-1}\sigma_1 E &= \sigma_1 & \sigma_1^{-1}\sigma_1\sigma_1 &= \sigma_1 \\ \sigma_3^{-1}\sigma_1\sigma_3 &= \sigma_2 & C_3^{-1}\sigma_1 C_3 &= \sigma_3 \\ \sigma_2^{-1}\sigma_1\sigma_2 &= \sigma_3 & (C_3^2)^{-1}\sigma_1 C_3^2 &= \sigma_2 \end{aligned} \tag{3.20}$$

and

$$[\sigma_1] = \{\sigma_1, \sigma_2, \sigma_3\}$$

We need not bother with finding the elements conjugate to σ_2 or σ_3, that is, the classes $[\sigma_2]$ and $[\sigma_3]$, since we know by the transitive property of conjugation that they will be identical to $[\sigma_1]$. Accordingly, we choose one of the remaining elements, say C_3,

$$\begin{aligned} E^{-1}C_3E &= C_3 & \sigma_1^{-1}C_3\sigma_1 &= C_3^2 \\ \sigma_3^{-1}C_3\sigma_3 &= C_3^2 & C_3^{-1}C_3C_3 &= C_3 \\ \sigma_2^{-1}C_3\sigma_2 &= C_3^2 & (C_3^2)^{-1}C_3C_3^2 &= C_3 \end{aligned} \tag{3.21}$$

Thus, $[C_3]=\{C_3, C_3^2\}$. We have exhausted all the elements of the group in terms of putting them into disjoint classes so that our job is completed. Thus, C_{3v} has the three classes: $\{E\}$, $\{\sigma_1, \sigma_2, \sigma_3\}$, and $\{C_3, C_3^2\}$. We note that the class orders 1, 3, and 2 are divisors of $^{o}G=6$ and that the orders of all elements in a given class are the same (1, 2, 3, respectively).

Problem

3.16 Show that for abelian groups every element is in a class by itself.

The relationship between elements in a class has a simple interpretation in terms of groups of operators and coordinate transformations. It is most readily visualized in terms of symmetry operations and we will approach it this way; the generalization to operators in general should be clear. Consider a right-handed coordinate system and two column vectors v_1, v_2 expressed in it. Suppose that v_1 and v_2 are related by some point symmetry operation A, which has the effect of changing v_1 into v_2.

$$Av_1 = v_2 \tag{3.22}$$

A could be, for example, a rotation or reflection inversion through the origin or some combination of these. For example, if A were "rotate by 90° clockwise looking down the z-axis" and if v_1 represented a unit vector in the y-direction of a right-handed coordinate system, v_2 would represent a unit vector in the x-direction. If we wished to express these two column vectors in a second coordinate system we would need to know the transformation operator U that would take us from the first (unprimed) to the second (primed) system

$$v_1 = Uv_1', \qquad v_2 = Uv_2' \tag{3.23}$$

Given this transformation it is readily seen that v_1' and v_2' are also related in the primed system by the operator $A' = U^{-1}AU$:

$$\begin{aligned} Av_1 &= v_2 \\ AUv_1' &= Uv_2' \\ (U^{-1}AU)v_1' &= v_2' \end{aligned} \tag{3.24}$$

$U^{-1}AU$, just as conjugation $s^{-1}gs$, is an example of what we call a *similarity* transformation. The operator $U^{-1}AU$ does in the primed system what A did in the unprimed system.

Consider then a group of symmetry operators and two operations A and B in the same class so that $A = C^{-1}BC$, for some $C \in G$. Then, the fact that A and B are conjugate indicates that the operation performed by B in one coordinate system is equivalent to the *type* performed by A in a second coordinate system where the two coordinate systems are related by another symmetry operation (the coordinate transformation) also in the group.

Consider C_{4v} and the two operations C_4 and C_4^3. C_4 is just our previous "rotate clockwise 90° looking down the z-axis" operator and likewise C_4^3 is a counterclockwise 90° rotation. In an ordinary right-handed coordinate system, C_4 would take a unit vector in the y-direction, ε_y into a unit vector in the x-direction ε_x. If we change to a second coordinate system where the transformation operator is, say, $\sigma_v^{(yz)}$, the vertical mirror plane in the yz-plane, we obtain a left-handed coordinate system. In this second system, to perform the same *type* of operation—that is, to take ε_y' into ε_x'—we must rotate counterclockwise by 90°; thus, in the transformed coordinate system C_4^3 does what C_4 did in the untransformed system. Therefore, C_4 and C_4^3 are conjugate and belong to the same class. If we study the situation for the operator C_2 we will find that there is no coordinate system available *in the set of group symmetry operations* that replaces the effect of C_2 by another operator. Hence in C_{4v} $\{C_2\}$ forms a class by itself. The remaining classes for C_{4v} are $\{E\}$, $\{\sigma_v, \sigma_v'\}$, and $\{\sigma_d, \sigma_d'\}$.

It is important to remember in the above example that one compares the *same type of process in different coordinate systems*; our example above was ε_y into ε_x in the primed and ε_y' into ε_x' in the unprimed systems. Looked at as vectors in a common coordinate system they result from different operations, just as C_4 and C_4^3 are obviously different symmetry operations. Those different symmetry operators are conjugate which yield the same *type* of process in coordinate systems related by another symmetry operator in the group.

A pictorial scheme for recognizing conjugate symmetry point group operations may be helpful. Construct an image (a picture) of a particular operation. See what happens to the image (the picture) when some operation of the group carries it into another image. The second image or picture obtained will be an image or picture of some other symmetry operation in the group. The two pictures or images so related by a symmetry operation in the group will correspond to elements that are conjugate to one another and therefore in the same class. By such a scheme we can fairly readily divide the group operations in the point groups into classes. For example, two vertical or diagonal reflections will be in the same class if the group contains a symmetry operation that moves one plane into the other. Two rotations about different axes each taken k times [C_n^k and $(C_n^k)'$ or S_n^k and $(S_n^k)'$] likewise will belong to the same class if there is a symmetry operation in the group that carries one n-fold axis into the other n-fold axis. For rotations about the same axis, the situation is as follows: C_n^k and C_n^{-k} are in the same class (but different for each k) if there is a plane of symmetry containing the C_n-axis or if there is a C_2 at a right angle to C_n. The same relation obtains for the improper rotations S_n^k and S_n^{-k}. Reflection in a horizontal plane (σ_h) and inversion, i, are clearly each in a class by themselves. This result is most easily seen by recognizing the fact that these two operations are unique among the point group symmetry operations; there is nothing else equivalent to them so that by conjugation they must be carried only into themselves. What this means in terms of the group algebra is that these two operations will commute with all other operations in the group.

It is clear that in classes similar elements are related. It is no accident, therefore, that rotations are conjugate to other rotations and planes to other planes; we never find a plane conjugate to a rotation or a rotation corresponding to one angle conjugate to a rotation corresponding to an angle of a different absolute magnitude.

The reader may find the review of vectors and matrices in Chapter 6 helpful in becoming more at ease with the geometrical interpretation of classes.

SOLUTIONS TO PROBLEMS

3.1 For $a \in G$ and $a^n = e$ (n = order of a), then $H = \{a, a^2, a^3, \ldots, a^n = e\} \subset G$ and $^\circ H = n$. Since $^\circ G \div {}^\circ H$ = an integer, the order of an element is a divisor of $^\circ G$.

3.2 If $^\circ G$ is prime, only trivial or improper subgroups can exist for in such a case only 1 and $^\circ G$ are divisors. Thus, for any element a of order n, $a^n = e$ only for $n = {}^\circ G$ so that G must be cyclic.

3.3 If H is of index 2 in G, then $G = H \cup (aH)$ for $a \notin H$ and also $G = H \cup (Ha)$ so clearly $aH = Ha$. Thus, any subgroup of index 2 must be normal. Stated another way, $G - H = aH = Ha$, an equality arising from the two preceding equalities.

3.4 In an abelian group *all* elements commute. Thus $aH = \{ae, ah_1, ah_2, \ldots\} = \{ea, h_1a, h_2a, \ldots\} = Ha$ for all a and H is normal.

3.5 $H_1 = \{E, \sigma_1\}$, $H_2 = \{E, C_3, C_3^2\}$

$\sigma_2 H_1 = \{\sigma_2, C_3^2\} = C_3^2 H_1 \qquad H_1\sigma_2 = \{\sigma_2, C_3\} = H_1 C_3$

$\sigma_3 H_1 = \{\sigma_3, C_3\} = C_3 H_1 \qquad H_1\sigma_3 = \{\sigma_3, C_3^2\} = H_1 C_3^2$

$\sigma_1 H_2 = \{\sigma_1, \sigma_2, \sigma_3\} = \sigma_2 H_2 = \sigma_3 H_2$

$H_2\sigma_1 = \{\sigma_1, \sigma_3, \sigma_2\} = H_2\sigma_2 = H_2\sigma_3$

H_1 is not normal while H_2, being of index 2, must be a normal subgroup.

3.6 Let a be the element of order 3 so that the group contains the subgroup $\{a, a^2, a^3 = e\}$. Let b be distinct from these elements so that

$$G = \{e, a, a^2, b, ba, ba^2\}$$

If $b^3 = e$, then b^2 must be one of the above six distinct elements. But by assumption $b \neq e, a, a^2$ and $b^2 = b$, $b^2 = ba$, or $b^2 = ba^2$ implies $b = e$, $b = a$, and $b = a^2$, respectively, which violates our assumption. Thus, b cannot be of order 3 and must be of order 2. Consider the product ab. Clearly $ab \neq e, a, b$, or a^2 so either $ab = ba$ or $ab = ba^2$. If $ab = ba$ then

$$(ab)^2 = (ab)ba = a^2 \qquad \text{(since } b^2 = e\text{)}$$

$$(ab)^3 = (ba)a^2 = b$$

$$(ab)^4 = (ab)b = a$$

$$(ab)^5 = (ba)a = ba^2$$

$$(ab)^6 = (ab)(ba^2) = e$$

and the element ab is of order 6. This cannot be so by our initial assumption that we are dealing with a *noncyclic* group. Thus the alternative, $ab = ba^2$, must hold. This is allowed since $(ab)^2 = (ab)(ba^2) = e$ is of order 2. Thus, the group table may be written as shown in Table 1.

Table 1

	e	a	a^2	b	ba	ba^2
e	e	a	a^2	b	ba	ba^2
a	a	a^2	e	ba^2	b	ba
a^2	a^2	e	a	ba	ba^2	b
b	b	ba	ba^2	e	a	a^2
ba	ba	ba^2	b	a^2	e	a
ba^2	ba^2	b	ba	a	a^2	e

(We may also show that there *must* be an element of order 3 present so that the above group plus the cyclic group of order 6 exhaust the structures for order 6 groups.)

3.7 For $g, g' \in G$, if $g^{-1}g' \in G$, then G is a group. Thus, consider the two elements of $s^{-1}Hs$:

$$s^{-1}h_i s, \qquad s^{-1}h_j s$$

Then,

$$\begin{aligned}(s^{-1}h_i s)^{-1}s^{-1}h_j s &= s^{-1}h_i^{-1}ss^{-1}h_j s \\ &= s^{-1}h_i^{-1}h_j s \\ &= s^{-1}h_k s \in H\end{aligned}$$

and $s^{-1}Hs$ is a group.

3.8 Problem 3.5 shows that the subgroups $\{E, \sigma_1\}$, $\{E, \sigma_2\}$, and $\{E, \sigma_3\}$ are not normal but that $\{E, C_3, C_3^2\}$, being of index 2, is. Thus, letting

$$H_2 = \{E, C_3, C_3^2\}$$
$$G = H_2 \cup H_2\sigma_1$$

we see that G/H_2 along with the (trivial) factor groups $\{E\}$ and C_{3v} are the factor groups of C_{3v}.

3.9 If H is normal $gh_ig^{-1} \in H$, all $g \in G$. Since all $g = hk$ (i.e., are of this form)

$$\begin{aligned}gh_ig^{-1} &= hkh_i(hk)^{-1} \\ &= hkh_ik^{-1}h^{-1} \\ &= hh_ih^{-1} \in H\end{aligned}$$

since h, k commute. In a like manner K is shown to be normal.

Suppose g is an element in the intersection of H and K. As our other element in the direct-product group, g can be expressed as hk in only one way; but if g is in $H \cap K$, $g \in H$ and $g \in K$ and $g = ge = eg$ can be written in two ways, except if $g = e$. Thus, $H \cap K$ can only contain e if the group is to be a direct-product group.

3.10 $C_{2h} = \{E, C_2, \sigma_h, i\}$ with proper subgroups

$$C_2 = \{E, C_2\}, \quad C_s = \{E, \sigma_h\}, \quad \text{and} \quad C_i = \{E, i\}$$

These are all normal subgroups since they are of index 2 in C_{2h}. Thus $C_{2h} = C_2 \times C_s = C_2 \times C_i = C_s \times C_i$. For $C_s \times C_i$ we see that

$$\begin{aligned}C_sC_i &= \text{product of } \{E, \sigma_h\}, \text{with } \{E, i\} \\ &= \{E, i, \sigma_h, \sigma_h i = C_2\} = C_{2h}\end{aligned}$$

3.11 C_{3v} has the proper subgroups $\{E, \sigma_1\}$, $\{E, \sigma_2\}$, $\{E, \sigma_3\}$, and $\{E, C_3, C_3^2\}$. Problem 3.5 shows that while the cyclic subgroup of order 3 is normal, the others are not. Since two normal subgroups are not present it is not possible to write C_{3v} as a direct product of its proper subgroups. Note that we do not consider the "trivial" use of the group G being the direct product of the improper or trivial subgroups $\{e\}$ and G itself.

3.12 a. S_6. The product elements formed from $C_3 = \{E, C_3, C_3^2\}$ and $C_i = \{E, i\}$ are E, $i = S_6^3$. $C_3 = S_6^2$, $C_3 i = S_6^5$, $C_3^2 = S_6^4$, and $C_3^2 i = S_6$.
b. If we consider the direct product $C_3 \times C_s$ we would have obtained the point group C_{3h}. $C_3 C_s = E$, σ_h, C_3, $C_3\sigma_h = S_3$, C_3^2, $C_3^2\sigma_h = S_3$.

3.13 Let $a = s^{-1}bs$ and $b = k^{-1}ck$. Then $a = s^{-1}bs = s^{-1}k^{-1}cks = (ks)^{-1}c(ks)$ and a and c are conjugate to each other.

3.14 For *all* $s \in G$, $s^{-1}es = es^{-1}s = e$ so that e is conjugate only to itself. This is why the statement was made in the text that except for the group $\{e\}$ of order 1, all groups of $^\circ G > 1$ cannot have classes of order $^\circ G$. Every group has the class $[e]$, and, since the classes span the group, the other classes must have fewer than $^\circ G$ elements and, thus, orders less than $^\circ G$.

The facts that $[e]$ is always a class in G and classes are disjoint show us that except for the class $[e]$ classes are not subgroups.

3.15 Let C_g contain the elements a, b, and let a be of order n: $a^n = e$. Then since a and b are conjugate,

$$\begin{aligned} b^n &= (k^{-1}ak)^n \\ &= k^{-1}akk^{-1}akk^{-1}ak \cdots (n \text{ times}) \\ &= k^{-1}a^n k = e \end{aligned}$$

and b is also of order n.

3.16 For an abelian group, $ab = ba$, all $a, b \in G$. Thus, $s^{-1}as = as^{-1}s = a$, all $a, s \in G$. Each element is conjugate only to itself and thus each element of an abelian group is in a class by itself.

4
Mapping and Groups

THE MAPPING OF SETS

The idea of *mapping* (or *correspondence*) is a basic and elementary concept in the theory of sets. It allows us to accurately express some very fundamental situations. Most of us will readily recognize many of the situations that mapping describes; the terminology involved will allow us greater conciseness in describing those situations.

The idea of mapping in group theory (i.e., the mapping of groups) is important in that it provides the mathematical bridge connecting abstract groups and their real (in the sense of being nonabstract) representations. The existence of this bridge assures us that the various theorems of abstract groups will hold true for their real representations. As we shall see in later sections, we are most concerned with sets of matrices $\Gamma = \{\mathbf{D}(g)\}$ that possess the same structure as the abstract group G ($g \in G$); the set (which is a group) $\{\mathbf{D}(g)\}$ is termed a *representation* and one speaks of a mapping between the set of abstract elements $g \in G$ and the set, Γ, whose elements are the matrices $\mathbf{D}(g)$.

Consider two sets A, B and their elements $a \in A$ and $b \in B$. A *mapping*, f, exists between A and B

$$f\colon A \to B \qquad \text{or} \qquad A \xrightarrow{f} B \tag{4.1}$$

if a rule (or law or formula) exists that connects or associates with every element $a \in A$ a uniquely assigned element $b \in B$. Figure 1 illustrates

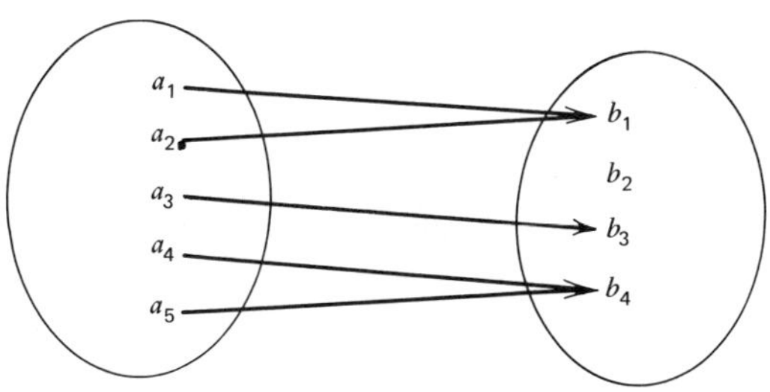

Figure 1

schematically a possible mapping $f: A \to B$ between two finite sets; the lines drawn connecting the elements represent the rule (law) that associates (connects) an element $a \in A$ to an element $b \in B$. The diagram illustrates the fact that in a mapping several elements of A may be mapped to the same element

$$b \in B(a_1, a_2 \xrightarrow{f} b_1)$$

and that there may be some elements $b \in B$ to which no elements of A are connected (b_2).

Mapping is directional in general so that the fact that $f: A \to B$ is a mapping does not allow us to presume that the reverse, $f: B \to A$, is also a mapping. The directional character of mapping is due to the fact that for each $a \in A$ there must be a *uniquely* determined element $b \in B$. Stated in terms of our schematic picture, one and only one line (the connection) must emanate from each element $a \in A$. Thus, in Figure 1, $B \to A$ is not a mapping because, for example, b_1 is connected to two elements of A and thus the connecting rule is not unique. Would $B \to A$ be a mapping if we erased a_2 and a_4 and their associated lines in the figure? The answer is no because *each* element of B must be connected to an element of A, and in our revised example b_2 is not connected to any element of A.

The element $b \in B$, associated with an element $a \in A$ in a mapping $f: A \to B$, is denoted as $f(a)$ $(b = f(a))$ and is called the *image* of a; the element or collection of elements which have b as their image is called the *fiber* of b and denoted as $f^{-1}(b)$. In our diagram $f(a_1) = b_1$ as well as $f(a_2) = b_1$; $f(a_3) = b_3$, and $f^{-1}(b_1) = \{a_1, a_2\}$ and $f^{-1}(b_3) = a_3$.

The collection of elements $f(a) \in B$ for all $a \in A$ is called the image of the set A and denoted by $f(A)$. If $f(A) = B$, we say that A is mapped *onto* B; that is, every $b \in B$ is an image of some one or more elements $a \in A$. If $f(A)$ does not span B $(f(A) \subset B$ in the nontrivial sense) we say that A is mapped *into* B. In this latter case, some one or more elements of b are not connected with an $a \in A$ or, using our set theory notation, $(B - f(A)) \cap B \neq \varnothing$. A is generally called the *domain* of f (the mapping) and B is called the *range* of f.

Often mappings possess a *multiplicity*. If the fiber of b [$f^{-1}(b)$, the collection of elements $a \in A$ having b as their image] consists of n elements $a \in A$ for every b involved in the mapping [i.e., all $b \in f(A)$], we speak of an n-to-one (or $n:1$) mapping. If the mapping $f: A \to B$ is 1:1 and "onto" then $B \to A$ is also a mapping and is the *inverse mapping* of f. A 1:1 and "onto" mapping of a set onto *itself* is called a *transformation;* if the set involved is finite, the transformation is called a *permutation.*

There are many common examples of mappings. The set of students in a classroom can be mapped onto the 2-element set {male, female}, "onto" provided there is at least one member of each sex present. The people in the classroom could be mapped into the set of integers where the rule of mappings could be, for example, the person's age in years or his weight in pounds. "$y = \cos x$" is an example of a mapping of a set into itself (real numbers) with the rule "take the cosine of"; here, unless we restrict the domain of x to be $\pm\pi/2$, an inverse mapping will not exist because the relation $x = \arccos y$ is one-to-many. A 1:1 mapping of the letters of the alphabet onto the first 26 positive integers is an example of a code; if the code is one that mixes up the letters of the alphabet in a 1:1 mapping, we have a permutation (a transformation involving a finite set). A trivial mapping is the *identity mapping* in which each element of a set is simply mapped onto itself.

We can have products of mappings. For example, given a set of elements $a \in A$ and two rules for mappings f_1 and f_2, we can produce the mapped set $f_2(f_1(A))$ by first performing f_1 and then f_2 or the mapped set $f_1(f_2(A))$, if we perform f_2 first and then f_1. The two results will not in general be the same; that is, products of mappings do not necessarily commute.

THE MAPPING OF GROUPS

Consider a mapping between two sets with structure, $f: A \to B$. If the same set of relations exist between the images of the $a_i \in A$, $f(a_i) \in B$, as exist between the $a_i \in A$, we say that the mapping *preserves the structure.* We are interested in groups so that we are interested in mappings that preserve *algebraic* structure.

Let A and B be two groups, and for the purpose of emphasizing a point let us use temporarily the composition notation "$\cdot$" and "$*$," where $\cdot$ and $*$ denote the composition laws for the groups A and B, respectively. We use this notation temporarily to emphasize that although A and B possess algebraic structure, their composition laws need not be the same in the considerations that follow. Thus "multiplication" in A, "$\cdot$," might be ordinary multiplication while the composition law in B, "$*$," might be addition modulo some integer. In A let $a_1 \cdot a_2 = a_3$; then if the mapping is to preserve the (algebraic) structure, it must be true that $f(a_1) * f(a_2) = f(a_3)$ where $f(a_i) \in B$. If this condition obtains, we speak of a *homomorphic mapping* or simply a *homomorphism.* If the homomorphism is 1:1 *and* onto, it is called an *isomorphic mapping* (iso = equal) or simply an *isomorphism.*

The test for homomorphism is more readily stated by noting that if $a_3 = a_1 a_2$ (we now drop the different composition law notation) then

$f(a_3) = f(a_1 a_2)$ so that

$$f(a_1)f(a_2) = f(a_1 a_2) \tag{4.2}$$

for any two $a_i \in A$. Keep in mind that the specific composition laws in the groups A, B need not be the same.

As an example of a homomorphic mapping consider our previous examples of the two cyclic groups of order 4: $A = \{1, i, -1, -i\}$ under ordinary multiplication, and $B = \{0, 1, 2, 3\}$ under addition modulo 4. Both groups have the same multiplication table (see problem 2.4) so that $f: A \to B$ is a homomorphism. The association between the elements of the two sets is clearly

$$\begin{aligned} 1 &\to 0 = f(1) \\ i &\to 1 = f(i) \\ -1 &\to 2 = f(-1) \\ -i &\to 3 = f(-i) \end{aligned} \tag{4.3}$$

This is clearly a 1:1 and onto homomorphism that is, an isomorphism; in this example A and B are isomorphic. Clearly $f: B \to A$ in this instance is also a valid mapping. Both groups are also obviously isomorphic to the abstract cyclic group of order four.

As a second example consider the two groups $G = \{1, i, -1, -i\}$ and $G' = \{1, -1\}$, both under ordinary multiplication. If we make the association (the mapping)

$$\begin{aligned} &G \xrightarrow{f} G' \\ &\{1, -1\} \to 1 = f(1) = f(-1) \\ &\{i, -i\} \to -1 = f(i) = f(-i) \end{aligned} \tag{4.4}$$

we obtain a 2:1 homomorphism. To check that the condition that must hold

$$f(g_1)f(g_2) = f(g_1 g_2) \tag{4.5}$$

is indeed satisfied, we note that

$$\begin{aligned} f(1)f(-1) &= (1)(1) = 1 = f(-1) = f((1)(-1)) \\ f(-1)f(i) &= (1)(-1) = -1 = f(-i) = f((-1)(i)) \end{aligned} \tag{4.6}$$

The other products can also be shown to satisfy the homomorphism condition.

Problems

4.1 Consider a set of square nonsingular matrices $\{\mathbf{M}_i\}$ which form a group under matrix multiplication. Show that the mapping $f(\mathbf{M}_i) = \det \mathbf{M}_i$ is a homomorphism.

4.2 Consider a group G of elements $g \in G$ and the mapping $f(g) = a^{-1}ga$ for some *fixed* choice of $a \in G$. Show that this conjugation is an isomorphism.

Homomorphic mappings possess a number of important and useful properties. We have already introduced the terms image and fiber. A particularly important fiber in the mapping of two groups $f\colon G \to G'$ is the fiber of e', the unit element of the image set $f(G) \subset G'$; the fiber of e', $f^{-1}(e')$ is called the *kernel* of the homomorphic mapping $f\colon G \to G'$. By definition it contains all those elements $g \in G$ which have $e' \in G'$ as their image: $K = \{k_i \in G \mid f(k_i) = e'\}$. In the first example above of homomorphic mappings the kernel is simply the unit element itself, 1, of the group $\{1, i, -1, -i\}$; in the second example the kernel is the *set* $\{1, -1\}$.

It should be clear that for a homomorphism $f\colon G \to G'$ the image set $f(G)$ must be a subgroup of G' since by the homomorphism condition the image elements $f(g)$ possess algebraic structure and satisfy the group conditions. (See problem 4.5 below.) It can be shown that the kernel not only is a subgroup of G but that it is a normal subgroup of G. Not only can we demonstrate that the kernel of a homomorphism is a normal subgroup but also that *only* normal subgroups can serve as the kernel of a mapping and indeed that *each* normal subgroup of G can serve as a kernel for a mapping.

We state finally that *all* fibers of homomorphic mappings have the same number of elements; that is, every homomorphic mapping has a *definite multiplicity*.

Problems

4.3 Show that $f(e) = e' \in G'$ in $f\colon G \to G'$.

4.4 Show that $f(g)^{-1} = f(g^{-1})$.

4.5 Show that $f(G)$ is a subgroup of G' for $f\colon G \to G'$.

4.6 Show that K is a normal subgroup of G where K is the kernel of the mapping $f\colon G \to G'$.

4.7 Show that a homomorphic mapping has a definite multiplicity.

FACTOR GROUPS AND HOMOMORPHISMS; THE ISOMORPHISM THEOREM

We spoke earlier of left and right quotient sets and normal (invariant) subgroups. Problem 4.2 showed that $f(g) = s^{-1}gs$ for some fixed $s \in G$ was an isomorphism. If we consider a subgroup $H \subset G$ and form the mapping $f(H) = s^{-1}Hs$ for $s \in G$ and *either* $s \in H$ or $s \notin H$ another isomorphism exists. If $s \in H$ the situation is just as before; if $s \notin H$ then, since the elements $s^{-1}hs \in G$, $f(H)$ is clearly a subgroup of G. $s^{-1}Hs$ is called the *subgroup conjugate to H by s.* If H is a normal subgroup $s^{-1}Hs = H$, *all* $s \in G$, and another way of expressing normality of a subgroup is to say that it is *self-conjugate.*

With every normal subgroup $N \subset G$ we can form a factor group G/N consisting of the symbolic elements $N, aN, bN, \ldots$, where N plays the role of the identity element in the factor group. The formation of the factor group G/N is clearly a homomorphism, $f\colon G \to G/N$ where N is the kernel of the mapping. As stated in the previous section, since only normal subgroups can serve as kernels of homomorphic mappings and each normal subgroup may serve as a kernel, we arrive at what is usually called the *isomorphism theorem:*

Every homomorphic mapping of G is isomorphic to some factor group of G

The isomorphism theorem shows us that knowledge of a group's normal subgroups and its factor groups is sufficient to determine all possible homomorphisms.

SOLUTIONS TO PROBLEMS

4.1 If $f(\mathbf{M}_i) = \det \mathbf{M}_i$ is a homomorphism, then $f(\mathbf{M}_i)f(\mathbf{M}_j) = f(\mathbf{M}_i\mathbf{M}_j)$ should hold. It does hold since $\det \mathbf{M}_i\mathbf{M}_j = \det \mathbf{M}_i \det \mathbf{M}_j$:

$$\begin{aligned} f(\mathbf{M}_i)f(\mathbf{M}_j) &= \det \mathbf{M}_i \det \mathbf{M}_j \\ &= \det \mathbf{M}_i\mathbf{M}_j \\ &= f(\mathbf{M}_i\mathbf{M}_j) \end{aligned}$$

4.2 This is a homomorphism since
$$\begin{aligned} f(g_1)f(g_2) &= s^{-1}g_1ss^{-1}g_2s \\ &= s^{-1}g_1g_2s \\ &= f(g_1g_2) \end{aligned}$$

It is an isomorphism since the image of each g_i is unique. If two elements of $g \in G$ had the same image,

$$s^{-1}g_1s = s^{-1}g_2s$$
$$g_1 = g_2$$

and the uniqueness property of set elements would be violated.

4.3 Since $f(g_1)f(g_2) = f(g_1g_2)$, then $f(g)f(e) = f(ge) = f(g)$ and clearly $f(e) = e'$.

4.4 $f(g)f(g^{-1}) = f(gg^{-1}) = f(e) = e'$ so that $f(g)^{-1} = f(g^{-1})$.

4.5 $f(G)$ is a group (a subgroup of G') if for $f(g_1), f(g_2) \in f(G) \subset G'$ also $f(g_1)^{-1}f(g_2) \in f(G)$. It follows that this is true since

$$\begin{aligned} f(g_1)^{-1}f(g_2) &= f(g_1^{-1})f(g_2) \\ &= f(g_1^{-1}g_2) \\ &= f(g_3) \in f(G) \end{aligned}$$

4.6 Let $k_i \in K$ where $K = f^{-1}(e')$. If $k_1 \in K$, then also $k_1^{-1} \in K$ since

$$f(k_1^{-1}) = f(k_1)^{-1} = (e')^{-1} = e'$$

K is a group (a subgroup of G) since for $k_1, k_2 \in K$ also $k_1^{-1}k_2 \in K$:

$$\begin{aligned} f(k_1^{-1}k_2) = f(k_1^{-1})f(k_2) = f(k_1)^{-1}f(k_2) &\\ = (e')^{-1}(e') &\\ = e' & \end{aligned}$$

To prove that K is a normal subgroup of G we must show that its left and right cosets are the same. Let $g \in G$ but $g \notin K$. Then $f(gk_i) = f(g)f(k_i) = f(g)e' = f(g)$ for all $k_i \in K$ and all members of the left coset of K by g have the same image, $f(g)$. It is also easily seen to be true that the elements of the right coset of K by g also have the same image $f(g)$. No other element of G can have the same image; for if $f(h) = f(g)$ for $h \notin gK$, $f(g^{-1})f(h) = f(g)^{-1}f(h) = e' = f(g^{-1}h)$ so that $g^{-1}h \in K$; however, this implies $h \in gK$, which is a contradiction. Thus each left (or right) coset of K has a distinct image. Since the left and right cosets of K by g have the *same* image, the left and right cosets of K are the same and K is normal.

4.7 Problem 4.6 shows that each left (or right) coset of K has a distinct image in G'. Since each coset has the same number of elements as K, each fiber $f^{-1}(g')$ for $g' \in G'$ is associated with a particular coset of K and has the same number of elements as every other fiber. Thus, a definite multiplicity exists for homomorphic mappings.

5
Permutation Groups

PERMUTATIONS

Cayley's theorem, derived below, tells us that every group of order n is isomorphic with some subgroup of the symmetric group S_n associated with the $n!$ permutations of n objects. It is obviously an important result, for it indicates that knowledge of the permutation group is sufficient to determine the structure of all groups. This is sufficient reason to introduce permutations and the symmetric groups at this point. There are, however, other good reasons for spending some time on the subject. Permutations in general are important in science as they relate to physical systems of identical and indistinguishable particles. Such systems—because of the indistinguishability of the particles contained therein—possess a permutation symmetry; the Hamiltonian of such a system is left unchanged if we permute in an arbitrary way the indices of the identical particles. That such permutation symmetry is physically important is clear from the division of particle statistics into the Bose–Einstein and Fermi–Dirac classifications. The wave functions of such systems must show *particular* symmetries with respect to permutations. Besides these two points, permutations present a particularly nice example of mapping. For these reasons we should have a basic working knowledge of the associated theory.

Consider an ordered array of n distinguishable objects. There are $n!$ different ways of ordering these objects. Thus, if we have the array of three letters a, b, c (or indexed symbols x_1, x_2, x_3) we can order them in the following $3! = 6$ ways:

$$\begin{array}{lll} abc\ (x_1x_2x_3) & bac\ (x_2x_1x_3) & cab\ (x_3x_1x_2) \\ acb\ (x_1x_3x_2) & bca\ (x_2x_3x_1) & cba\ (x_3x_2x_1) \end{array}$$

Each arrangement represents a permutation of some basic array, say the array "abc" ($x_1x_2x_3$). We would say that the array "abc" is the result of the *identity* permutation acting on abc whereas bac results from interchanging the first and second elements while leaving the third alone. We can express the rearrangement of an array from any one arrangement to any other by the symbol

$$\begin{pmatrix} 1 & 2 & 3 & \cdots & n \\ p_1 & p_2 & p_3 & \cdots & p_n \end{pmatrix}$$

which we interpret as meaning object #1 (element of the array with index 1) is changed to object p_1 (index p_1), object #2 (index #2) is changed to object p_2, and so forth. Thus

$$\begin{pmatrix} 123 \\ 231 \end{pmatrix}$$

would take *abc* ($x_1x_2x_3$) into *bca* ($x_2x_3x_1$) or *bac* ($x_2x_1x_3$) into *cba* ($x_3x_2x_1$) while

$$\begin{pmatrix} 123 \\ 321 \end{pmatrix}$$

would, for example, take *abc* into *cba*, or take $x_1x_3x_2$ into $x_3x_1x_2$. Note that it is not the ordering of the vertical pairs in the permutation symbol that is important but simply what the pairs are. Thus,

$$\begin{pmatrix} 132 \\ 213 \end{pmatrix} \quad \text{and} \quad \begin{pmatrix} 321 \\ 132 \end{pmatrix}$$

represent the same operation as

$$\begin{pmatrix} 123 \\ 231 \end{pmatrix}$$

A simpler notation suggests itself which is best explained by means of examples. Instead of writing the two line symbols as above we would write

$$\begin{pmatrix} 123 \\ 312 \end{pmatrix} \Rightarrow (132)$$

and translate the right-hand symbol in terms of the left-hand symbol as follows: object (index) #1 changes into object (index) #3, object (index) #3 changes into object (index) #2, object (index) #2 changes into (index) #1. Starting at the left of the symbol we simply have the previous vertical pairs now as horizontal overlapping pairs. The brackets are used to indicate the termination of a sequence; that is, the end element in a bracket (cycle) is changed to the object (index) indicated by the first element in the bracket; in the above example, $2 \rightarrow 1$ or

$$(1 \curvearrowright 3 \curvearrowright 2), \quad 2 \rightarrow 1$$

The symbol

$$\begin{pmatrix} 123 \\ 321 \end{pmatrix}$$

becomes (13) (2) where we read the second bracket as $2 \to 2$, which obviously means that the object (index) #2 of the array is left unchanged. The identity operation in this example would be simply (1)(2)(3). A bracketed group of numbers is called a *cycle;* in expressing a permutation the order of *disjoint* cycles (cycles containing no common numbers) is immaterial:

$$(13)(2) = (2)(13) \tag{5.1}$$

as is the *cyclic order* within a cycle:

$$(123) = (312) = (231) \tag{5.2}$$

Any permutation can be written as the product of mutually disjoint cycles. Some exercises will help make the notation clear.

Problem

5.1 Transcribe the following permutations into products of disjoint cycles.

a. $\begin{pmatrix} 12345 \\ 24351 \end{pmatrix}$

b. $\begin{pmatrix} 1234567 \\ 4213765 \end{pmatrix}$

c. $\begin{pmatrix} 3142 \\ 4213 \end{pmatrix}$

Often in discussing a permutation of *degree* n (i.e., one of the $n!$ permutations of n objects), if a particular element in the array is not affected by the permutation it is simply left out of the permutation symbol. Thus, for the degree 3 permutation (13)(2) we could simply write (13), for the degree 7 permutation (143)(2)(57)(6) simply (143)(57), implying in all cases that what does not appear is left unchanged. We cannot do this, of course, for the identity permutation and will use for it the symbol e

$$e = (1)(2)(3) \cdots (n) \tag{5.3}$$

THE SYMMETRIC GROUP

It should be clear that for an array of n objects there are precisely $n!$ arrangements and precisely $n!$ permutations that create these arrangements from some one basic array or that take any one arrangement into all the others. The $n!$ permutations form a group, called the *symmetric group* S_n of order $n!$. S_n is a group because the set is finite and stable and the associative law is obeyed; the product of two permutations must be another permutation in the set.

In order to see the group nature of S_n we need to look at the product of permutations. Thus, the product of (13)(24) (on the left) and (1243) (on the right) is

$$(13)(24)(1243) = (14)(2)(3) \tag{5.4}$$

which we obtain by following the progress of any one index as it is modified by the various operations. Thus, for the first index the first operation is $1 \to 2$, the second operation is $2 \to 4$, with the *net* effect that $1 \to 4$; we continue on until this cycle is complete noting that for #4 the sequence is $4 \to 3$ and $3 \to 1$ for a net effect of $4 \to 1$, thus closing this cycle. For the second index the sequence is $2 \to 4$, $4 \to 2$ with the *net* effect $2 \to 2$; a similar result obtains for index #3 of $3 \to 3$. Note that the order of the permutations is important; they will not in general commute.

$$\begin{aligned}(1243)(13)(24) &= (1)(23)(4) \\ &\neq (13)(24)(1243)\end{aligned} \tag{5.5}$$

It should be easy to see that the inverse of a permutation consisting of disjoint cycles is the permutation consisting of the same cycles in reverse cyclic order; that is,

$$(123 \cdots n)^{-1} = (n \cdots 321) \tag{5.6}$$

since

$$\begin{aligned}(123 \cdots n)(n \cdots 321) &= (n \cdots 321)(123 \cdots n) \\ &= (1)(2) \cdots (n) \\ &= e\end{aligned} \tag{5.7}$$

Thus

$$[(14)(2)(3)]^{-1} = (41)(2)(3) = (14)(2)(3) \tag{5.8}$$

and it is clear that this group element, being its own inverse, is of order 2. Another example of the inverse of a permutation is

$$[(1437)(528)]^{-1} = (7341)(825) \tag{5.9}$$

Problems

5.2 Find the elements of the symmetric group S_3 by associating with each different permutation of three objects a permutation of degree 3. Construct the multiplication table for S_3.

5.3 Show that S_3 and C_3 are isomorphic.

CAYLEY'S THEOREM

Although we will make additional comments on the symmetric groups S_n and its subgroups, we have enough background at this point to derive

Cayley's theorem and will now do so. Consider a group G of elements $e, a, b, c, \ldots \in G$ and its multiplication table

G	e	a	b	c	…
e	e	a	b	c	..
a	a	a^2	ab	ac	…
b	b	ba	b^2	bc	…
c	c	…	…	…	…
.	.				
.	.				
.	.				

where we reserve the symbol e for the identity element and arbitrarily use it to label the leading row and column of the group table. Consider the *ordered* sequence of elements labeling the columns as the basic array of $^{\circ}G$ objects. The rearrangement theorem tells us that each row contains the same set of symbols but in a different order. That is, each row is obtained from the "basic row" by a suitable *permutation* of the order of appearance of the group element in the column heading array. Denote a given row as a row vector; thus, for example, the third row above is

$$(b, ba, b^2, bc, \ldots)$$

which can be obtained by the "operation" of b on the ordered sequence $(e, a, b, c, \ldots)$.

$$b(e, a, b, c, \ldots) = (b, ba, b^2, bc, \ldots) \tag{5.10}$$

In terms of the reordering of the sequence of elements b has the effect of the permutation

$$\begin{pmatrix} e & a & b & \cdots & \\ b & ba & b^2 & bc & \cdots \end{pmatrix}$$

where now we are using the equivalent ordering of letters (but with $e =$ "#1") instead of an ordering of numbers. Thus, b in the abstract group G is mapped to the permutation expressed

$$\begin{pmatrix} 1 & 2 & 3 & \cdots \\ n_{b1} & n_{b2} & n_{b3} & \cdots \end{pmatrix}$$

where we have numbered the group elements according to their order of appearance in the group column heading (the "basic array") and the numbers $n_{b1}, n_{b2}, \ldots$ represent the reordered sequence obtained by the operation of b on the basic sequence $(e, a, b, \ldots)$. To illustrate, consider the $^{\circ}G=3$ group and the mapping to certain permutations of degree 3.

	e	a	b
e	e	a	b
a	a	b	e
b	b	e	a

1	2	3
1	2	3
2	3	1
3	1	2

$e \to (1)(2)(3)$
$a \to (123)$
$b \to (132)$

The intermediate table is constructed to illustrate the manner of associating the permutations with the group elements.

We have performed a mapping, but it remains to demonstrate that it is a homomorphism. If the set of permutations which is the image set of G is a group, then the mapping will be an isomorphic mapping since all the permutations are obviously different (since the rows of the multiplication table are all different). If the mapping is a homomorphism

$$f(b)f(c)=f(bc) \tag{5.11}$$

and the permutation associated with the product bc must equal the product of the permutations of $f(b)$ and $f(c)$ associated with the elements b, c. Let $(a_1, a_2, a_3, \ldots)$ be the basic ordered array so that

$$\begin{aligned} b &\xrightarrow{f} p_b = \begin{pmatrix} a_1 & a_2 & a_3 & \cdots \\ ba_1 & ba_2 & ba_3 & \cdots \end{pmatrix} \\ c &\xrightarrow{f} p_c = \begin{pmatrix} a_1 & a_2 & a_3 & \cdots \\ ca_1 & ca_2 & ca_3 & \cdots \end{pmatrix} \end{aligned} \tag{5.12}$$

and

$$bc \xrightarrow{f} p_{bc} = \begin{pmatrix} a_1 & a_2 & a_3 & \cdots \\ bca_1 & bca_2 & bca_3 & \cdots \end{pmatrix} \tag{5.13}$$

Recall now the fact that the relative ordering of the symbols in the two-line permutation symbol is not important as long as vertical pairs are maintained together. Thus, we may, for example, write

$$p_b = \begin{pmatrix} a_1 & a_2 & \cdots \\ ba_1 & ba_2 & \cdots \end{pmatrix} = \begin{pmatrix} ca_1 & ca_2 & \cdots \\ bca_1 & bca_2 & \cdots \end{pmatrix} \tag{5.14}$$

so that

$$\begin{aligned} p_b p_c &= \begin{pmatrix} ca_1 & ca_2 & \cdots \\ bca_1 & bca_2 & \cdots \end{pmatrix} \begin{pmatrix} a_1 & a_2 & a_3 & \cdots \\ ca_1 & ca_2 & ca_3 & \cdots \end{pmatrix} \\ &= \begin{pmatrix} a_1 & a_2 & a_3 & \cdots \\ bca_1 & bca_2 & bca_3 & \cdots \end{pmatrix} \\ &= p_{bc} \end{aligned} \tag{5.15}$$

and the homomorphism is demonstrated. Now the image set of permutations are some, but obviously not all, of the permutations of degree n, and they form a group; since all permutations of degree n are elements of the symmetric group S_n of order $n!$, the image set is clearly a subgroup of order n of the parent group S_n. We have proven *Cayley's theorem* which states:

> All *groups of order n are isomorphic with a subgroup of the symmetric group* S_n.

Thus, the importance of permutation groups: If we know all there is to know about the permutation groups we know the possible structures of all groups.

Problem

5.4 Find the subgroups of S_4 that are isomorphic to the two abstract groups of order 4.

The entire set of manipulations above could have been performed with the columns of the multiplications table instead of the rows. The symmetry of the group table in this respect tells us the results would be identically the same.

The type of permutations involved in the mapping of G of order n onto a subgroup of S_n is a special kind. They are permutations that leave no element of an array unchanged; this must be so since the permutations are based on the uniqueness of the row (or column) order structure in the group multiplication table. Such permutations are called *regular permutations*. Regular permutations are characterized by the fact that their cycle structure is such that only cycles of the same length (same number of elements in a cycle) can occur. In the examples of problem 5.4 the permutations had one 4-element cycle, or two 2-element cycles, or, for $e =$ the identity element, four 1-element cycles. The cycle structure characteristic follows because the permutations are part of a subgroup so that all the powers must belong to the subgroup. If one has a cycle of k elements then that cycle raised to the kth power is equivalent to the unit operator for the elements involved.

$$(123 \cdots k)^k = (1)(2)(3) \cdots (k) \tag{5.16}$$

Thus, if two cycles of different length occur on a permutation it is possible to find a power of that permutation that resolves one or more cycles into unit cycles while leaving others unchanged in cycle length. This would mean a permutation in which some symbols are left unchanged, and, except for the obvious case of the identity permutation, this is not allowed in the class of regular permutations.

THE ALTERNATING GROUP

There is one important subgroup of the symmetric group which bears mentioning. It is called the *alternating group* A_n and is a subgroup of order $n!/2$ in S_n. To understand the alternating group we need to introduce two further ideas concerning permutations. A 2-cycle or cycle consisting of just two symbols—such as (ab) or (13)—is called a *transposition;* in such a cycle one merely exchanges or transposes the positions of the objects with the indicated symbols. Now any cycle can be written as the product of transpositions; thus, for example,

$$\begin{aligned}(123) &= (13)(12)\\ (1234) &= (14)(13)(12)\end{aligned} \tag{5.17}$$

or, in general, for a n-cycle (a cycle of n symbols),

$$(123\cdots n) = (1n)\cdots(13)(12) \tag{5.18}$$

The cycle on n symbols can be written as the product of $n-1$ transpositions. Accordingly, every permutation can be written as a product of transpositions; if the permutation has k cycles of length $l_1, l_2, \ldots, l_k$ (where several l_i may be the same), it can be written as a product of $(l_1-1)(l_2-1)\cdots(l_k-1)$ transpositions. If the number of such transpositions $(l_1-1)+(l_2-1)+\cdots+(l_k-1)$, is even, the permutation is said to be *even;* if the number of transpositions is odd, the permutation is called *odd.*

Problem

5.5 Verify in the following examples of degree 5 permutations that the odd or even character of a permutation is changed from one to the other when multiplied by a single transposition.
   a. (12) and (345)
   b. (12) and (143)(25)
   c. (12) and (15243)

Problem 5.5 illustrates the basis for the statement that the product of two odd permutations is even as is the product of two even permutations, while the product of an even and an odd permutation is odd. If we let P_j^o and P_k^e represent the jth odd and kth even permutations then our algebra is

$$\begin{aligned}&P_j^oP_i^o = P_k^e\\ &P_k^eP_q^e = P_m^e\\ &P_j^oP_k^e = P_r^o; \qquad P_k^eP_j^o = P_s^o\end{aligned} \tag{5.19}$$

where the last line reminds us that in general permutations will not commute. Thus, S_n may be divided into two sets of permutations characterized by their even or odd nature.

Problem

5.6 Show that the number of even and odd permutations in S_n is the same and equal to $n!/2$.

The set of $n!/2$ even permutations is stable and is therefore a subgroup of S_n. The set of $n!/2$ odd permutations is not stable and does not contain an identity element and is not a subgroup. The subgroup of even permutations of degree n of order $n!/2$ is called the *alternating group* and denoted by A_n. Since A_n is of index 2 in S_n, it is a normal subgroup and can be used as the kernel of an isomorphic mapping to a group of order 2. This is nothing more than the factor group S_n/A_n of order 2 which is the simple and unique abstract group of order 2.

To merely hint at the importance of the alternating group in physical theory, recall the characteristics of systems of identical particles and the quantum mechanical wave functions that represent them. Such systems possess permutation symmetry; being identical and indistinguishable particles, the Hamiltonian of the system must remain unchanged regardless of how the particle numbers are interchanged (permuted). This permutation symmetry is expressed in a very special way which depends upon the parity—the evenness or oddness—of the permutations involved. If the particles are of integral spin (Bose–Einstein particles or bosons) any permutation of the particle labeling leaves the wave function of the system unchanged. On the other hand, if the particles are of half-integral spin (Fermi–Dirac particles or fermions) the wave function remains unchanged under an *even* permutation and changes sign under an *odd* permutation. The symmetric group and its representations play an important role in our description of the physical world.

CLASS STRUCTURE AND PERMUTATIONS

Because of an important relation between class structure in a group and certain of its particular representations (the irreducible representations) we close this section on the symmetric group with a few comments without proof on the decomposition of S_n into classes. We can first of all show that all permutations in the same class (conjugation class) have identical disjoint cycle structure and that each class is characterized by its disjoint cycle structure alone. It follows that the classes in S_n can be characterized by the various decompositions of n into a sum of positive integers (the order is unimportant) and the number of classes is given by the total number of ways of performing such a decomposition. Thus, for

S_4, $n=4$ and

$$\begin{aligned} 4 &= 4 \\ &= 3+1 \\ &= 2+2 \\ &= 2+1+1 \\ &= 1+1+1+1 \end{aligned} \tag{5.20}$$

There are five classes: one containing all unit cycles (obviously e, always in a class by itself); one containing only a single transposition $(2+1+1)$; one consisting of elements containing pairs of transpositions $(2+2)$; one consisting of elements containing only one 3-cycle $(3+1)$; and one consisting of only a 4-cycle permutation.

Problems

5.7 Find the classes of S_3 and verify that the disjoint cycle structure is characteristic of the class. Use the isomorphism $f: S_3 \rightarrow C_{3v}$ to find the elements belonging to the various classes.

5.8 Indicate the number of classes and their cycle structure for S_2 and S_5.

It can also be shown that the number of elements in a class characterized by n_{l_i} cycles of length l_i is given by

$$\frac{n!}{\prod_i l_i^{n_{l_i}}(n_{l_i}!)}$$

Thus in S_4 the number of elements in the various classes are given by

Class	*Number of elements*
4	$\frac{4!}{4(1!)}=6$
3+1	$\frac{4!}{3^1(1!)1^1(1!)}=8$
2+2	$\frac{4!}{2^2(2!)}=3$
2+1+1	$\frac{4!}{2^1(1!)1^2(2!)}=6$
1+1+1+1	$\frac{4!}{1^4(4!)}=1$

(5.21)

Clearly the total number of elements is $24 = 4!$, the order of the group,

$$\sum_i n_{1_i} = n! \tag{5.22}$$

Problems

5.9 Use the above formula to find the number of elements in the classes of S_3 and S_5.

5.10 Show that the parity of permutations in the same class is the same.

SOLUTIONS TO PROBLEMS

5.1 a. $\begin{pmatrix}12345\\24351\end{pmatrix} = (1245)(3)$

b. $\begin{pmatrix}1234567\\4213765\end{pmatrix} = (143)(2)(57)(6)$

c. $\begin{pmatrix}3142\\4213\end{pmatrix} = (1234)$

5.2 The 3! arrays and their corresponding permutation operators (based on the "basic" array) are

abc (basic array)	$e = (1)(2)(3)$
acb	$= (23)$
bac	$= (12)$
bca	$= (123)$
cab	$= (132)$
cba	$= (13)$

The multiplication table is

	e	(12)	(13)	(23)	(123)	(132)
e	e	(12)	(13)	(23)	(123)	(132)
(12)	(12)	e	(132)	(123)	(23)	(13)
(13)	(13)	(123)	e	(132)	(12)	(23)
(23)	(23)	(132)	(123)	e	(13)	(12)
(123)	(123)	(13)	(23)	(12)	(132)	e
(132)	(132)	(23)	(12)	(13)	e	(123)

5.3 Consider the diagram from problem 2.5 where the vertices associated with the vertical planes σ_1, σ_2, and σ_3 are labeled as 1, 2, and 3, respectively. Associate with each point group operation of

C_{3v} the appropriate permutation of the number labels to obtain Figure 1.

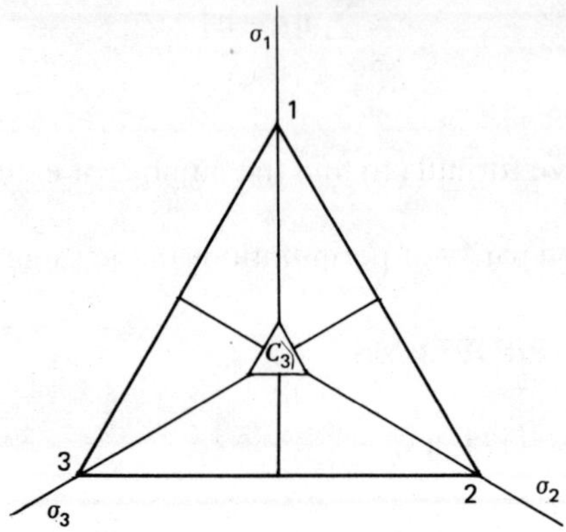

Figure 1

$$E = e \qquad \sigma_1 = (23)$$
$$\sigma_3 = (12) \qquad C_3^2 = (123)$$
$$\sigma_2 = (13) \qquad C_3 = (132)$$

This is the isomorphic mapping f: $C_{3v} \rightarrow S_3$. It is easily verified that the structure of the two multiplication tables is identical. That the use of this approach is limited is shown by the relatively small number of point groups whose order is the factorial of an integer and where the integral number of points exhibits the full symmetry of the group; thus, the latter part of this last statement shows why C_6 is not isomorphic to S_3 (besides the fact that C_6 is, of course, not isomorphic to C_{3v}).

5.4 Consider first the cyclic order 4 group.

C_4	e	a	b	c	
e	e	a	b	c	$e \rightarrow (1)(2)(3)(4)$
a	a	b	c	e	$a \rightarrow (1234)$
b	b	c	e	a	$b \rightarrow (13)(24)$
c	c	e	a	b	$c \rightarrow (1432)$

For the 4 group:

V	e	a	b	c	
e	e	a	b	c	$e \rightarrow (1)(2)(3)(4)$
a	a	e	c	b	$a \rightarrow (12)(34)$
b	b	c	e	a	$b \rightarrow (13)(24)$
c	c	b	a	e	$c \rightarrow (14)(23)$

5.5 a. (345) is clearly even.
(12) · (345) = (12)(1)(2)(345) = (12)(345) is clearly odd.
b. (143)(25) is odd.
(12)(143)(25) = (14325) is even.
c. (15243) is even.
(12)(15243) = (15)(243) is odd.

5.6 Suppose there are n_1 even permutations and $n_2 = n! - n_1$ odd permutations in S_n which consists of $n!$ elements. Consider the group table with the even permutations listed first (the identity e is even) and some row in the table associated with the permutation P_2^o.

S_n	$P_1^e = e \quad P_2^e \cdots \quad P_{n_1}^e$	$P_1^o \quad P_2^o \quad \cdots \quad P_{n_2}^o$
⋮		
P_2^o	$A = \{P_2^o P_k^e\}$, odd	$B = \{P_2^o P_j^o\}$, even
⋮		

The set A consisting of the n_1 products of P_2^o with all the P_k^e is odd and distinct from the set B which must be even. The sets A and B must span the group; thus A, consisting of n_1 odd permutations, must contain all odd permutations and $n_1 = n_2 = n!/2$.

5.7 From the isomorphism between C_{3v} and S_3 we know the classes of S_3 to be

$$[E] \xrightarrow{f} e = (1)(2)(3)$$

$$[\sigma_1] = \{\sigma_1, \sigma_2, \sigma_3\} \xrightarrow{f} \{(12), (13), (23)\}$$

$$[C_3] = \{C_3, C_3^2\} \xrightarrow{f} \{(123), (132)\}$$

$$\begin{aligned} 3 &= 3 \\ &= 2+1 \\ &= 1+1+1 \end{aligned}$$

and S_3 consists of 3 classes with the above cycle structure.

5.8 S_2: $n = 2 = 2$ 2 classes

$$= 1+1$$

S_5: $n = 5 = 5$ 7 classes

$$\begin{aligned} &= 4+1 \\ &= 3+2 \\ &= 3+1+1 \\ &= 2+2+1 \\ &= 2+1+1+1 \\ &= 1+1+1+1+1 \end{aligned}$$

5.9 S_3: $n = 3$ $\qquad n_c = \frac{3!}{3} = 2$

$= 2+1 \qquad = \frac{3!}{2} = 3$

$= 1+1+1 \qquad = \frac{3!}{3!} = 1$

S_5: $n = 5 \qquad n_c = 5!/5 \qquad = 24$

$= 4+1 \qquad = 5!/4 \qquad = 30$

$= 3+2 \qquad = 5!/3 \cdot 2 \qquad = 20$

$= 3+1+1 \qquad = 5!/3(2!) \qquad = 20$

$= 2+2+1 \qquad = 5!/2^2(2!) \qquad = 15$

$= 2+1+1+1 \qquad = 5!/2(3!) \qquad = 10$

$= 1+1+1+1+1 \qquad\qquad = 1$

$120 = 5!$

5.10 Since the disjoint cycle structure of the inverse of a permutation is the same as the permutation itself, odd permutations will be conjugate only to certain other odd permutations and even permutations to only certain other even permutations. Using P_j^o and P_j^e to symbolically represent odd and even permutations

$$(P_i^e)^{-1}P_j^o P_i^e \text{ is odd}$$

$$(P_i^o)^{-1}P_j^o P_i^o \text{ is odd}$$

and

$$(P_i^e)^{-1}P_j^e P_i^e \text{ is even}$$

$$(P_i^o)^{-1}P_j^e P_i^o \text{ is even}$$

so that elements in the same conjugation class are all either even or odd.

6
Vectors, Matrices, and Transformations: A Review

In the following sections we discuss the construction of representations in terms of matrices and transformations in a vector space. The ideas of vector spaces, basis functions (basis vectors), and transformations are so useful in quantum theory and group theoretical applications that we digress briefly here to review the fundamental ideas involved.

In what follows we speak of vector spaces of arbitrary dimensionality. It often proves helpful to visualize the concepts presented in two- or three-dimensional space with which we have some direct physical experience and then to realize the mathematical extension to spaces of higher dimension. Also, recall that we can speak of functions, basis functions, and function expansion by treating such a function space as a vector space. A few brief words at the end of this section will remind us of this important generalization.

An n-dimensional vector space is defined by a set of n linearly independent basic vectors $\vec{u}_1, \vec{u}_2, \ldots, \vec{u}_n$; the set $\{\vec{u}_j\}$ is called a (the) *basis* or a (the) *basis set*. Any arbitrary vector $\vec{v}$ in the vector space can be expanded in terms of the basis set

$$\vec{v} = \sum_j c_j \vec{u}_j \tag{6.1}$$

The basis set is said to *span* the vector space.

Given the basis set describing a vector space, that is, defining a coordinate system, the vector $\vec{v}$ can be specified by giving the set of components of the basis $\{\vec{u}_j\}$, in the form of a column matrix (column vector)

$$\begin{pmatrix} c_1 \\ c_2 \\ \cdot \\ \cdot \\ \cdot \\ c_n \end{pmatrix} = \mathbf{c} \tag{6.2}$$

To completely specify $\vec{v}$ *both* $\mathbf{c}$ and the basis must be specified. One often speaks of *linear* vector spaces by which is meant that vector addition is associative and commutative and that multiplication of a vector by a complex scalar is distributive and associative.

Any set of n linearly independent vectors can serve as a basis for an n-dimensional vector space, and therefore the choice of basis for a particular vector space is not unique. Figure 1 represents equally valid basis sets for an ordinary two-dimensional space. There are, of course, many more bases, because the condition for linear independence in two dimensions is simply that the two vectors not be colinear.

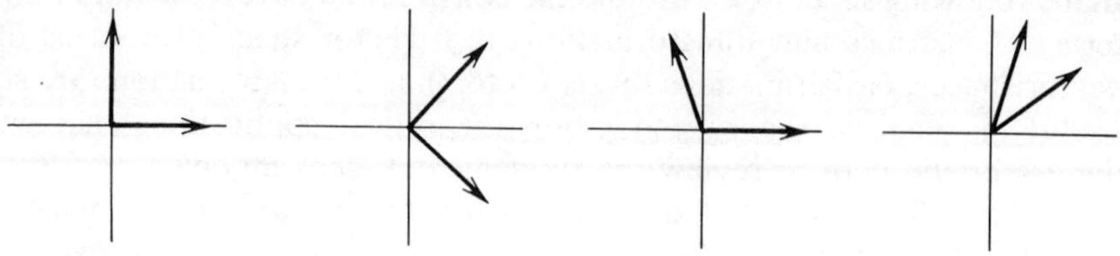

Figure 1

The column vector **v** representing the vector $\vec{v}$ in one basis $\{\vec{u}_j\}$ will have generally a different appearance (different components) in a second basis $\{\vec{u}'_j\}$. If

$$\vec{v} = \sum_j c_j \vec{u}_j = \sum_k c'_k \vec{u}'_k$$

$$\mathbf{c} = \begin{pmatrix} c_1 \\ c_2 \\ \cdot \\ \cdot \\ \cdot \\ c_n \end{pmatrix}; \quad \mathbf{c}' = \begin{pmatrix} c'_1 \\ c'_2 \\ \cdot \\ \cdot \\ \cdot \\ c_{n'} \end{pmatrix} \tag{6.3}$$

Although the vector $\vec{v}$ has an intrinsic significance, the *components* of $\vec{v}$ clearly depend on the coordinate system (basis set) in which it is described. Let the two-coordinate systems be related by the transformation

$$\vec{u}'_k = \sum_j \vec{u}_j \mathbf{T}_{jk} \tag{6.4}$$

where **T** is the *transformation matrix*. Then since

$$\vec{v} = \sum_j c_j \vec{u}_j = \sum_k c'_k \vec{u}'_k$$

$$= \sum_{kj} c'_k \vec{u}_j \mathbf{T}_{jk} \tag{6.5}$$

and since the basis vectors $\{\vec{u}_j\}$ (and $\{\vec{u}'_k\}$) are independent

$$c_j = \sum_k \mathbf{T}_{jk} c'_k \tag{6.6}$$

or in matrix notation the representation of $\vec{v}$ in the two-coordinate systems is expressed by

$$\mathbf{c} = \mathbf{T}\mathbf{c}' \tag{6.7}$$

With each pair of vectors $\vec{v}_1$ and $\vec{v}_2$ in the vector space there is associated a complex number $(\vec{v}_1, \vec{v}_2)$ called the *scalar product* of $\vec{v}_1$ and $\vec{v}_2$. (Any vector space in which such a scalar product is defined is called a *unitary vector space.*) Since we allow for complex components $(\vec{v}_1, \vec{v}_2) = (\vec{v}_2, \vec{v}_1)^*$ and in terms of the basis set $\{\vec{u}_j\}$

$$(\vec{v}_1, \vec{v}_2) = \sum_{ij} c_i^{(1)*} c_j^{(2)} (\vec{u}_i, \vec{u}_j) \tag{6.8}$$

The numbers $(\vec{u}_i, \vec{u}_j) \equiv m_{ij} = m_{ji}^*$ represent the scalar products of the basis vectors describing the coordinate system and are characteristic of the basis. They form a matrix $\mathbf{M}$ called the *metric matrix* or *metric tensor* of the system. Thus,

$$(\vec{v}_1, \vec{v}_2) = \mathbf{c}^{(1)\dagger} \mathbf{M} \mathbf{c}^{(2)} \tag{6.9}$$

where $\mathbf{c}^{(1)\dagger}$ is the *adjoint* of $\mathbf{c}^{(1)}$ and is the (complex conjugate) row vector associated with $c^{(1)}$

$$\mathbf{c}^{(1)\dagger} = (c_1^{(1)*} c_2^{(1)*} \cdots c_n^{(1)*}) \tag{6.10}$$

In general, the adjoint of a matrix $\mathbf{A}$ is defined by

$$(\mathbf{A}^\dagger)_{ij} = (\mathbf{A})_{ji}^* \tag{6.11}$$

The symmetry of the metric matrix shows us that $\mathbf{M}^\dagger = \mathbf{M}$: such a matrix is said to be *self-adjoint* or *hermitian.* In quantum mechanics hermitian matrix representations are required for dynamical variables because hermitian matrices have real eigenvalues.

The *length* or *norm* of a vector $\vec{v}$ is given by its scalar product with itself

$$(\vec{v}_1, \vec{v}_1) = \mathbf{c}^\dagger \mathbf{M} \mathbf{c} \tag{6.12}$$

Two vectors $\vec{v}_1$ and $\vec{v}_2$ are said to be *orthogonal* if their scalar product is zero

$$\mathbf{c}^{(1)\dagger} \mathbf{M} \mathbf{c}^{(2)} = 0 \tag{6.13}$$

It is clearly convenient to work in a basis in which the basis vectors are all orthogonal and normalized to unity, $(\vec{u}_j, \vec{u}_k) = \delta_{jk}$, for then $\mathbf{M}$ becomes diagonal and equal to the unit matrix, and scalar products take on the simpler form

$$(\vec{v}_1, \vec{v}_2) = \mathbf{c}^{(1)\dagger}\mathbf{c}^{(2)} \tag{6.14}$$

Such bases are called orthonormal *bases* and correspond in ordinary space to having all the coordinate axes at right angles to each other. Orthonormality of the basis set is not necessary but is often convenient. As in any physical problem the choice of the coordinate system cannot affect the physical result; we can, however, by a judicious choice of basis, often make the physics of the situation more transparent or accelerate the process of determining the desired result.

Suppose two bases are related by

$$\vec{u}'_j = \sum_k \vec{u}_k T_{kj} \tag{6.15}$$

If both bases are orthonormal, then

$$(\vec{u}'_j, \vec{u}'_k) = \delta_{jk} = \sum_{n,m} (\vec{u}_n, \vec{u}_m) T^*_{nj} T_{mk}$$

$$= \sum_n T^*_{nj} T_{nk}$$

$$= \sum_n T^\dagger_{jn} T_{nk} \tag{6.16}$$

or, writing **1** for the unit matrix of the vector space

$$\mathbf{1} = \mathbf{T}^\dagger \mathbf{T} \tag{6.17}$$

or

$$\mathbf{T}^{-1} = \mathbf{T}^\dagger \tag{6.18}$$

A matrix whose inverse is equal to its adjoint is said to be *unitary*. The above has shown us that two orthonormal bases are related by a unitary transformation.

Suppose two vectors $\vec{a}$, $\vec{b}$ in the orthonormal basis $\{\vec{u}_j\}$ are related by

$$\mathbf{a} = \mathbf{R}\mathbf{b} \tag{6.19}$$

where **a**, **b** are column vectors (vector components in $\{\vec{u}_j\}$), and **R** is a matrix. We can imagine that **R** represents the result of some operator, $\hat{O}_R$, operating on the vector **b** to give the vector **a**; **R** is then the matrix representation of the operator $\hat{O}_R$ in the basis $\{\vec{u}_j\}$. Just as a vector has a significance independent of the coordinate system in which it is represented, so too does the operator $\hat{O}_R$. But, again, just as the components of a vector depend upon the basis, so too does the matrix representation of $\hat{O}_R$. If there is a second orthonormal basis $\{\vec{u}'_j\}$ related to the first such that

$$\mathbf{a} = \mathbf{T}\mathbf{a}' \qquad \mathbf{b} = \mathbf{T}\mathbf{b}' \tag{6.20}$$

then clearly

$$\begin{aligned} \mathbf{a} &= \mathbf{T}\mathbf{a}' = \mathbf{R}\mathbf{b} = \mathbf{R}\mathbf{T}\mathbf{b}' \\ \mathbf{a}' &= \mathbf{T}^{-1}\mathbf{R}\mathbf{T}\mathbf{b}' = \mathbf{R}'\mathbf{b}' \\ \mathbf{R}' &= \mathbf{T}^{-1}\mathbf{R}\mathbf{T} \end{aligned} \tag{6.21}$$

where the last equation gives us the form that the matrix representation of $\hat{O}_R$ takes in the second basis. Any two matrices related by such an equation are said to be related by a *similarity transformation.* We should note the obvious resemblance of the similarity transformation and conjugation classes of groups. Our earlier discussion of the geometrical interpretation of classes and transformations between coordinate systems should also be recalled.

In the discussion above we have been speaking of vectors, basis vectors, and vector expansions. The words functions, basis functions, and function expansion may be substituted because we can discuss a function space in exactly the same terms. The scalar product for function space will correspond to the integral

$$(\vec{u}_j, \vec{u}_k) \rightarrow \int u_j^* u_k \, d\tau \equiv \langle u_j \mid u_k \rangle \tag{6.22}$$

integrated over the appropriate space. Functions are orthogonal in the sense that the integral corresponding to the scalar product vanishes, and a normalized function is simply one whose integral with itself is unity

$$\langle u_j \mid u_j \rangle = \int u_j^* u_j \, d\tau = 1 \tag{6.23}$$

7
Representations and Their Construction

THE DEFINITION OF REPRESENTATIONS

Some of our concrete examples of groups have included multiplication of complex numbers, addition to a given modulus, and multiplication of permutation operators. In our discussion of mapping we mentioned the mapping of a group onto a set of matrices with matrix multiplication being the law of composition. All of these are examples of *representations* or *realizations* of groups in that they are specific, concrete, nonabstract sets which possess the group properties. Because of the great importance of matrices, the term *representation* has come to basically imply a *matrix* representation, and we shall use it as such, keeping in mind that a more general definition encompassing all nonabstract realizations of groups exists. Thus, *a representation* Γ *of an abstract group* G *of elements* $g_i \in G$ *is a group of matrices* $\mathbf{D}(g_i)$ *homomorphic to* G.

Γ is a set of matrices, one each ($\mathbf{D}(g_i)$) for each element (g_i) of the group, the composition law being ordinary matrix multiplication. The various matrices in Γ may or may not be different from each other. If all the matrices are different the representation is said to be *faithful*. A number of other properties of the matrix set Γ are seen by considering the homomorphism conditions $f(s)f(t)=f(st)$ or, using the matrix symbol $\mathbf{D}$ for the mapping,

$$\begin{aligned} &f\colon G \to \Gamma = \{\mathbf{D}(g_i)\} \\ &\mathbf{D}(s)\mathbf{D}(t) = \mathbf{D}(st), \quad \text{all } s, \quad t \in G \end{aligned} \tag{7.1}$$

Since both $\mathbf{D}(s)\mathbf{D}(t)$ and $\mathbf{D}(t)\mathbf{D}(s)$ must belong to Γ, the matrices must be square. If Γ consists of $n \times n$ matrices we speak of n as being the *dimension* of the representation. Since (for $e =$ identity element in G)

$$\mathbf{D}(s)\mathbf{D}(e) = \mathbf{D}(se) = \mathbf{D}(s) \tag{7.2}$$

it is clear that

$$\mathbf{D}(e) = \mathbf{1} \tag{7.3}$$

where $\mathbf{1}$ is the unit matrix of dimension n. We proved earlier that, in general,

$$f(s^{-1}) = [f(s)]^{-1} \tag{7.4}$$

Applied to the matrix group this gives

$$\mathbf{D}(s^{-1}) = [\mathbf{D}(s)]^{-1} \tag{7.5}$$

Thus, the matrix inverse of each $\mathbf{D}(s)$ must exist; since a nonvanishing determinant is required for a matrix inverse to exist each $\mathbf{D}(s)$ must be nonsingular (which means that its determinant must not vanish).

Problem

7.1 Verify that

a. $$\Gamma = \left\{ \begin{pmatrix} 1 & 0 \\ 0 & 1 \end{pmatrix}, \begin{pmatrix} \sqrt{3}/2 & 1/2 \\ 1/2 & -\sqrt{3}/2 \end{pmatrix} \right\}$$

is isomorphic to C_2.

b. $$\Gamma = \left\{ \begin{pmatrix} 1 & 0 & 0 \\ 0 & 1 & 0 \\ 0 & 0 & 1 \end{pmatrix}, \begin{pmatrix} 0 & 0 & 1 \\ 1 & 0 & 0 \\ 0 & 1 & 0 \end{pmatrix}, \begin{pmatrix} 0 & 1 & 0 \\ 0 & 0 & 1 \\ 1 & 0 & 0 \end{pmatrix} \right\}$$

is isomorphic to the abstract group of order 3.

c. the mapping of C_3

$$E, C_3, C_3^2 \xrightarrow{f} (1) = 1$$

$$\sigma_1, \sigma_2, \sigma_3 \xrightarrow{f} (-1) = -1$$

(where (1) and (−1) are the positive and negative one-dimensional unit matrices) gives a valid, nonfaithful, one-dimensional representation of C_{3v}.

Certain types of representatives are very easy to find. Remember that a representation results from a mapping and that any homomorphic image of a group G is isomorphic with some factor group G/N of G where N is a normal (invariant) subgroup and serves as the kernel (maps to the identity element) of the mapping. One trivial factor group of G is G/G or the unit group. All elements must map to the same matrix which must obviously be the matrix representing e, that is, $\mathbf{1}$ the unit matrix. We thus arrive at what are called the *identity* or *trivial* representations of G where all elements map to the unit matrix of whatever dimensionality we like. Although trivial, this example shows that the representations of a group are unlimited. Later on we shall see that with each finite group are associated a definite finite number of "fundamental" representations called irreducible representations.

One-dimensional representations are particularly easy to work with since a one-dimensional matrix is simply a number. We immediately note that $\mathbf{D}=0$ cannot occur as an element and that if we are dealing with a group of finite order the absolute magnitude of each element must be unity. That is to say, since the group is finite, the order of each element must be finite and such that its order n satisfies $g^n = e$; for a one-dimensional matrix this means that $\mathbf{D}(g)$ of order n must be equal to the nth root of unity and is of absolute magnitude 1. The following are examples of valid one-dimensional representations of C_2:

$$(1) \quad E \to 1$$
$$C_2 \to 1$$
$$(2) \quad E \to 1$$
$$C_2 \to -1$$

One-dimensional representations of cyclic groups are particularly easy to write down. Let s be the generator of the cyclic group $G(C_n)$ of order n. Since $s^n = e$, s can be represented by the nth root of unity or any integral power of it.

Problem

7.2 Construct the three distinct one-dimensional representations of C_3.

LINEAR TRANSFORMATIONS

Consider a vector space V_n spanned by the basis set $\{\phi_j\}$ of n basis vectors. As we approach the application of group theory in quantum mechanical systems we shall think more and more of *basis functions* and *function spaces* and shall use the ideas of vectors and functions interchangeably. To aid in becoming familiar with the usage we use ϕ_j now rather than $\hat{u}_j$ although the concepts are identical.

Suppose that with each element $g \in G$ for some group we can associate an operator $\hat{g}$ that effects a transformation of any given vector in the space to some other vector still in the space; that is, the space V_n is *stable* under the operations of the group. We may define the various operators according to their action on the basis vectors of V_n

$$\hat{g}\phi_i = \sum_{k=1}^{n} \phi_k D_{ki}(g) \in V_n, \qquad \text{all } g \in G \tag{7.6}$$

where $D_{ki}(g)$ is the ki-element of the matrix representation of the element (operator) g and the concluding part of the equation reminds us of the necessary stability of the space under all $g \in G$. *Given* then that such a vector space is available and *given* that an operator $\hat{g}$ can be associated with each element g of the group in question, it is true that the set of matrices $\Gamma = \{\mathbf{D}(g)\}$ is a representation of the group G. The dimension of the representation is n, the dimension of the vector space. The examples discussed below will show how the above "given's" are obtained.

It should be remarked that the actual construction of representations is not particularly important in *using* group theory for the important fundamental (irreducible) representations are known and we need simply refer to appropriate tables when in need. The important aspect of studying the construction of representations is to note their structure and to make clear the important physical connection with operators and the transformations of functions. The physical application of the theory starts from considerations of how functions transform.

The fact that Γ is a representation of G is easily shown by showing that the homomorphism condition obtains. Thus,

$$\begin{aligned} \hat{s}\hat{t}\phi_i &= \hat{s}(\hat{t}\phi_i) = \hat{s}\sum_k \phi_k D_{ki}(t) \\ &= \sum_{k,j} \phi_j D_{jk}(s) D_{ki}(t) \\ &= \sum_j \phi_j [\mathbf{D}(s)\mathbf{D}(t)]_{ji} \end{aligned} \tag{7.7}$$

and since, letting $a = st$,

$$\hat{a}\phi_i = \hat{s}\hat{t}\phi_i = \sum_j \phi_j D_{ji}(st) \tag{7.8}$$

Comparison of the two results shows clearly that the homomorphism condition obtains

$$\mathbf{D}(st) = \mathbf{D}(s)\mathbf{D}(t) \tag{7.9}$$

As we have previously indicated, $\mathbf{D}(e) = \mathbf{1}$, the unit matrix in n dimensions.

What we have shown then is that a set of linear transformations Γ of a vector space is a representation of a group G if Γ is a homomorphic image of G. The vector space V_n is called the *representation space* or *carrier space* of the representation Γ. The process of constructing a representation consists in finding a basis for a representation (choosing a coordinate system) and defining the operations of the group elements in the vector space.

We have already seen how to construct some one-dimensional representations. Let us consider two examples involving spaces of higher order. Consider the group C_{3v}—a favorite example—and the two-dimensional space defined in the figure below. The appropriate point group symmetry elements are also defined in Figure 1. The unit vectors $\vec{\varepsilon}_1$ and $\vec{\varepsilon}_2$ lie in the reflection planes defined by σ_1 and σ_2 and are not orthogonal. We shall form a representation by finding how the basis vectors $\vec{\varepsilon}_1$ and $\vec{\varepsilon}_2$ transform under the operations of the groups. Positive rotations are taken to be clockwise looking at the figure; thus, $\vec{\varepsilon}_1$ can be rotated into $\vec{\varepsilon}_2$ by a positive rotation of 120° (the operation C_3) or by a negative rotation of −240° (C_3^{-2}).

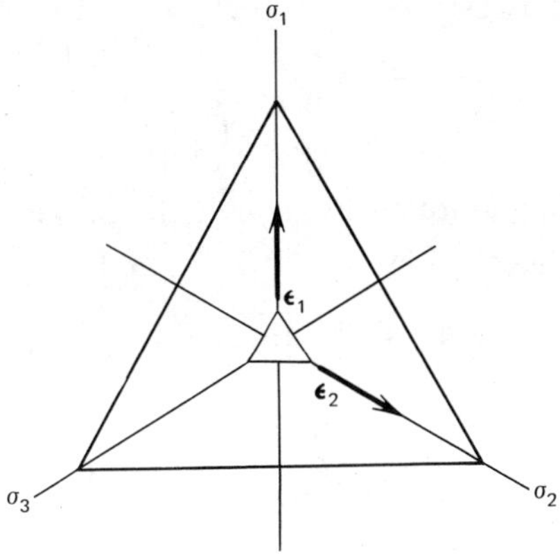

Figure 1

The reader should not worry about the fact that we are using three-dimensional symmetry operations and a two-dimensional vector space. Obviously the mirror planes require another dimension to "exist" as mirror planes, just as the threefold axis, extends in a direction perpendicular to the plane defined by $\vec{\varepsilon}_1$ and $\vec{\varepsilon}_2$. All we need be concerned with is that the operations of the group elements are definable on the basis set (in the space) in question and that operations on a vector in the space result in another vector still in the space (stability). This is certainly true in our example; any vector lying in the plane perpendicular to the C_3 axis remains in the plane for all the operations of C_{3v}. It should be clear that a two-dimensional vector space whose plane is not perpendicular to the threefold axis would not satisfy this criterion; that is, it would not be stable under the operations of the group. Thus, when considering three-dimensional symmetry operations, the allowable carrier spaces of dimensionally less than three cannot be chosen arbitrarily.

To find the two-dimensional representation we simply have to find the matrix representations of the C_{3v} point group operations. Take, for example, σ_1:

$$\begin{aligned} \hat{\sigma}_1\vec{\varepsilon}_1 &= \vec{\varepsilon}_1 \qquad & D_{11} &= 1, \qquad & D_{21} &= 0 \\ \hat{\sigma}_1\vec{\varepsilon}_2 &= -\vec{\varepsilon}_1 - \vec{\varepsilon}_2 \qquad & D_{12} &= -1, \qquad & D_{22} &= -1 \end{aligned} \tag{7.10}$$

or

$$D(\sigma_1) = \begin{pmatrix} 1 & -1 \\ 0 & -1 \end{pmatrix} \tag{7.11}$$

Note that $\vec{\varepsilon}_1+\vec{\varepsilon}_2$ is also a unit vector since $\vec{\varepsilon}_1 \cdot \vec{\varepsilon}_2=-\frac{1}{2}$. Looked at another way, for an arbitrary vector $\vec{v}=\alpha\vec{\varepsilon}_1+\beta\vec{\varepsilon}_2$

$$\hat{\sigma}_1(\alpha\vec{\varepsilon}_1+\beta\vec{\varepsilon}_2)=\alpha\vec{\varepsilon}_1+\beta(-\vec{\varepsilon}_1-\vec{\varepsilon}_2)$$
$$=(\alpha-\beta)\vec{\varepsilon}_1+(-\beta)\vec{\varepsilon}_2 \tag{7.12}$$

or

$$\mathbf{D}(\sigma_1)\begin{pmatrix}\alpha\\ \beta\end{pmatrix}=\begin{pmatrix}\alpha-\beta\\ -\beta\end{pmatrix} \tag{7.13}$$

which again, perhaps more easily, shows the proper form for $\mathbf{D}(\sigma_1)$. For σ_3, $\hat{\sigma}_3\vec{\varepsilon}_1=\vec{\varepsilon}_2$ and $\hat{\sigma}_3\vec{\varepsilon}_2=\vec{\varepsilon}_1$, so

$$\mathbf{D}(\sigma_3)\begin{pmatrix}\alpha\\ \beta\end{pmatrix}=\begin{pmatrix}\beta\\ \alpha\end{pmatrix}\Rightarrow\mathbf{D}(\sigma_3)=\begin{pmatrix}0 & 1\\ 1 & 0\end{pmatrix} \tag{7.14}$$

For C_3 we find

$$\mathbf{D}(C_3)\begin{pmatrix}\alpha\\ \beta\end{pmatrix}=\begin{pmatrix}-\beta\\ \alpha-\beta\end{pmatrix}\Rightarrow\mathbf{D}(C_3)=\begin{pmatrix}0 & -1\\ 1 & -1\end{pmatrix} \tag{7.15}$$

Since $C_3\sigma_1=\sigma_3$ we verify our results by observing that

$$\mathbf{D}(C_3)\mathbf{D}(\sigma_1)=\begin{pmatrix}0 & -1\\ 1 & -1\end{pmatrix}\begin{pmatrix}1 & -1\\ 0 & -1\end{pmatrix}=\begin{pmatrix}0 & 1\\ 1 & 0\end{pmatrix}=\mathbf{D}(\sigma_3) \tag{7.16}$$

Problem

7.3 Determine the remaining matrices for the above representation of C_{3v}.

As a second example we will consider C_{3v} again but in the guise of its isomorphic image group S_3, the symmetric group on three objects. Our vector space is a three-dimensional one (so we are going to obtain a three-dimensional representation) where the permutation operations associated with S_3 when operating on a row vector of basis vectors will be defined to permute the indices of the vector considered to be ordered initially from left to right in the row vector representation. Thus,

$$(132)\ (x_1x_2x_3)=(x_3x_1x_2)$$

and

$$(12)\ (x_1x_2x_3)=(x_2x_1x_3)$$

for example. It is readily apparent that to achieve the above transformations

$$\mathbf{D}(132)=\begin{pmatrix}0 & 1 & 0\\ 0 & 0 & 1\\ 1 & 0 & 0\end{pmatrix}$$

and

$$\mathbf{D}(12)=\begin{pmatrix}0 & 1 & 0\\ 1 & 0 & 0\\ 0 & 0 & 1\end{pmatrix} \tag{7.17}$$

where we would, for example, write as usual $(132)(x_1x_2x_3) = (x_1x_2x_3)\,\mathbf{D}(132)$.

Problem

7.4 Find the matrix representations for (123), (23), and (13) for the case in consideration.

We may have noted by this time that there is a very simple way to write down this type of representation. Note that if we associate the column vector $\begin{pmatrix}1\\0\\0\end{pmatrix}$ with "1", $\begin{pmatrix}0\\1\\0\end{pmatrix}$ with "2", and $\begin{pmatrix}0\\0\\1\end{pmatrix}$ with "3", the matrix associated with a given permutation is constructed by placing these column vectors from left to right to form a matrix in the permuted order as indicated by the operation. Thus, $e = (1)(2)(3)$ says column vector #1 to column vector #1, column vector #2 to vector #2, vector #3 to #3, and we get

$$\mathbf{D}(e) = \begin{pmatrix}1 & 0 & 0\\0 & 1 & 0\\0 & 0 & 1\end{pmatrix} \tag{7.18}$$

D(132) is constructed by reading the operation as column vector #1 changes into #3, vector #3 changes into #2, and vector #2 changes into #1, to obtain just what we derived above

$$\mathbf{D}(132) = \begin{pmatrix}0 & 1 & 0\\0 & 0 & 1\\1 & 0 & 0\end{pmatrix} \tag{7.19}$$

Problem

7.5 Find the matrix representations for S_2 and for the elements (1423), (123), and (13) (24) of S_4 using the simple technique just described.

Our observation shows us that *faithful* n-dimensional representations of this type for S_n, any n, can readily be written down. Since by Cayley's theorem all groups of order n are isomorphic to some subgroup of S_n, we are guaranteed of being able to construct representations of this type for any group.

Problem

7.6 Use Cayley's theorem and our simple method of constructing representations of permutation groups to construct a faithful three-dimensional representation of the abstract group of order 3.

THE REGULAR REPRESENTATION

The *regular representation* Γ_R of a group G is a particularly important one, since as we shall presently see it "contains" all the fundamental or irreducible representations of the group G. We introduce it here because it is constructed by just the means we have employed for permutation groups. Its relation to Cayley's theorem and regular permutations (and, thus, its name) will be evident.

Cayley's theorem was proved by demonstrating the isomorphic mapping between the elements of a group G of order n and a subgroup of permutations of the group S_n of order $n!$ The permutations involved were the so-called regular permutations, permutations that left no object in its original position in an ordered array. The permutations were identified readily from the multiplication table by considering the row labeling the columns of the group multiplication table as the basic ordered array and associating with each element labeling the rows that permutation operator which provided the permutation of the basic array as seen in that particular row. Thus, for C_3,

C_3	e	a	a^2
e	e	a	a^2
a	a	a^2	e
a^2	a^2	e	a

$$e \xrightarrow{f} (1)(2)(3)$$
$$a \xrightarrow{f} (123)$$
$$a^2 \xrightarrow{f} (132)$$

and using our method for constructing matrix representations for permutations we find, just as in problem 7.6,

$$\mathbf{D}^R(e) = \begin{pmatrix} 1 & 0 & 0 \\ 0 & 1 & 0 \\ 0 & 0 & 1 \end{pmatrix}; \qquad \mathbf{D}^R(a) = \begin{pmatrix} 0 & 0 & 1 \\ 1 & 0 & 0 \\ 0 & 1 & 0 \end{pmatrix};$$
$$\mathbf{D}^R(a^2) = \begin{pmatrix} 0 & 1 & 0 \\ 0 & 0 & 1 \\ 1 & 0 & 0 \end{pmatrix} \tag{7.20}$$

Thus, for any group G by so mapping the elements to their respective regular permutations and constructing the associated matrices according to our simple method, we generate the regular representation for that group. That is, the regular representation results from the $^{\circ}G$-dimensional matrix representation of the subgroup of regular permutation operator associated with the $^{\circ}G$ elements of the group G.

Problem

7.7 Find the regular representation of V, the noncyclic group of order 4.

The regular representation is faithful to the group G so that the study of Γ_R will reveal all the group theoretical properties of G. Because of this and because of its important relation to the irreducible representations of G, it is sometimes referred to as the *master representation.*

RELATED REPRESENTATIONS

We have seen how to construct representations by choosing carrier spaces and defining the operations associated with the group in question. Earlier we constructed some one-dimensional representations and most recently the regular representation was introduced. In this section we shall see that certain representations are "related" to others so that knowledge of these relationships allows us to construct still more representations of a group.

In the example we looked at in which the two-dimensional space of the plane perpendicular to the C_3-axis of the point group C_{3v} was the carrier space for a representation of C_{3v} we choose, somewhat arbitrarily, two basis vectors making an angle of 120° with each other. This was a *convenient* choice since the group operation of C_{3v} takes these vectors into *simple* linear combinations of each other. It was an *arbitrary* choice, however, since we could have chosen any two noncolinear unit vectors as a basis for a representation. For example, we might very well have first thought of a Cartesian system with the basis vectors at right angles to each other, an orthogonal basis. Since the two basis sets may be related by some transformation, the general properties of transformations tell us that matrices in one coordinate system can be simply related to those in the other coordinate system: In particular, if

$$\mathbf{D}(s)\mathbf{v} = \mathbf{v}_s \tag{7.21}$$

represents the action of the matrix representation of the group element $s \in G$, $\mathbf{D}(s) \in \Gamma$, on some (column) vector $\mathbf{v}$, and if the transformation connecting the two bases (coordinate systems) is $\mathbf{v} = \mathbf{U}\mathbf{v}'$, we have

$$\begin{aligned} \mathbf{D}(s)\mathbf{U}\mathbf{v}' &= \mathbf{U}\mathbf{v}'_s \\ \mathbf{U}^{-1}\mathbf{D}(s)\mathbf{U}\mathbf{v}' &= \mathbf{v}'_s \end{aligned} \tag{7.22}$$

and in the transformed system the matrix representation of $s \in G$ becomes $\mathbf{U}^{-1}\mathbf{D}(s)\mathbf{U}$. it follows readily that the set of transformed matrices $\Gamma' = \{\mathbf{U}^{-1}\mathbf{D}(s)\mathbf{U}\}$ is also a representation of the group G.

Problem

7.8 Use the homomorphism condition to show that the mapping $f(\mathbf{D}(s)) \equiv \mathbf{D}'(s) = \mathbf{U}^{-1}\mathbf{D}(s)\mathbf{U}$ is indeed a homomorphic mapping.

Accordingly, transforming from one basis set to another (within the same space, clearly) produces another representation. The new representation is new in form (appearance) but is obviously related to the old one by a similarity transformation. Two representations related by a similarity transformation are said to be *equivalent representations* and, although different in appearance, are considered to be basically the same or equivalent.

Another example of related representations is provided by the case in which a representation Γ is complex; that is, the matrices $\mathbf{D}(s) \in \Gamma$ have complex matrix elements. If $\Gamma = \{\mathbf{D}(s)\}$ is a representation, then $\Gamma^* = \{\mathbf{D}^*(s)\}$ is also a representation.

Problem

7.9 Show that Γ^* is a representation of G if Γ is.

THE ADDITION OF REPRESENTATIONS

A very important type of representation is concerned with what is termed the addition of representations. A square matrix is said to be *blocked* (or in block form) if it consists of smaller matrices arranged along the main diagonal (one can generalize to nonsquare matrices, but we will only consider square matrices). Thus, letting x represent a generally nonzero matrix element, consider the matrices

$$\left(\begin{array}{cc:c:ccc} x & x & & & & \mathbf{0} \\ x & x & & & & \\ \hdashline & & x & & & \\ \hdashline & & & x & x & x \\ & & & x & x & x \\ \mathbf{0} & & & x & x & x \end{array}\right), \quad \left(\begin{array}{cc:c:c:cc} x & x & & & & \mathbf{0} \\ x & x & & & & \\ \hdashline & & x & & & \\ \hdashline & & & x & & \\ \hdashline & & & & x & x \\ \mathbf{0} & & & & x & x \end{array}\right) \tag{7.23}$$

These are examples of blocked 6×6 matrices where the large zeros indicate that all elements outside of the dashed lines are definitely zero. In the first case the matrix blocks into one 2×2, a 1×1, and one 3×3 submatrix, while in the second example the blocking consists of two 2×2's and two 1×1's. Actually, all the "x's" need not be nonzero; there may be zeros in the blocked areas so long as they do not result in additional blocking. We can symbolically indicate the blocking of a matrix by

$$\mathbf{A} = \begin{pmatrix} \mathbf{A}_1 & & & & \mathbf{0} \\ & \mathbf{A}_2 & & & \\ & & \cdot & & \\ & & & \cdot & \\ & & & & \cdot \\ \mathbf{0} & & & & \mathbf{A}_n \end{pmatrix} \tag{7.24}$$

where $\mathbf{A}_1, \mathbf{A}_2, \ldots, \mathbf{A}_n$ are the smaller matrices. The $\mathbf{A}_i$ may all be different or all the same or any intermediate combination; if $\mathbf{A}_i$ has dimension d_i then $\sum_{i=1}^{n} d_i$ is the dimension of $\mathbf{A}$ itself. If two matrices $\mathbf{A}$ and $\mathbf{B}$ are blocked in exactly the same way—that is, $\mathbf{A}_i$ has the same dimension of $\mathbf{B}_i$ for all i, 1 to n, then the matrix product $\mathbf{AB}$ takes on the simple form

$$\begin{pmatrix} \mathbf{A}_1 & & & & 0 \\ & \mathbf{A}_2 & & & \\ & & \cdot & & \\ & & & \cdot & \\ 0 & & & & \mathbf{A}_n \end{pmatrix} \begin{pmatrix} \mathbf{B}_1 & & & & 0 \\ & \mathbf{B}_2 & & & \\ & & \cdot & & \\ & & & \cdot & \\ 0 & & & & \mathbf{B}_n \end{pmatrix} = \begin{pmatrix} \mathbf{A}_1\mathbf{B}_1 & & & & 0 \\ & \mathbf{A}_2\mathbf{B}_2 & & & \\ & & \cdot & & \\ & & & \cdot & \\ 0 & & & & \mathbf{A}_n\mathbf{B}_n \end{pmatrix} \qquad (7.24)$$

where $\mathbf{A}_i\mathbf{B}_i$ is the matrix product of dimension d_i of the smaller $\mathbf{A}_i$ and $\mathbf{B}_i$ matrices. The product $\mathbf{AB}$ is also blocked in just the same way as are $\mathbf{A}$ and $\mathbf{B}$.

Consider now two representations. $\Gamma_1 = \{\mathbf{D}^{(1)}(s)\}$ of dimension d_1 and $\Gamma_2 = \{\mathbf{D}^{(2)}(s)\}$ of dimension d_2 of a group G. Then, by the properties of blocked matrices, the following matrix equation is true:

$$\begin{pmatrix} \mathbf{D}^{(1)}(s) & 0 \\ 0 & \mathbf{D}^{(2)}(s) \end{pmatrix} \begin{pmatrix} \mathbf{D}^{(1)}(t) & 0 \\ 0 & \mathbf{D}^{(2)}(t) \end{pmatrix} = \begin{pmatrix} \mathbf{D}^{(1)}(st) & 0 \\ 0 & \mathbf{D}^{(2)}(st) \end{pmatrix} \qquad (7.25)$$

Accordingly, the set of matrices (all $s \in G$)

$$\Gamma_3 = \begin{pmatrix} \mathbf{D}^{(1)}(s) & 0 \\ 0 & \mathbf{D}^{(2)}(s) \end{pmatrix} \qquad (7.26)$$

is also a representation of G. Γ_3 results from the "addition" (direct sum) of the representations Γ_1 and Γ_2 in the above sense of forming a set of blocked matrices (matrix direct sum). One writes symbolically

$$\Gamma_3 = \Gamma_1 \oplus \Gamma_2 \qquad (7.27)$$

The dimension of Γ_3 is clearly $d_1 + d_2$, the sum of the dimensions of the representation Γ_1 and Γ_2.

The extension of this process of "adding" representations to form new representations should be clear. If $\Gamma_1, \Gamma_2, \ldots, \Gamma_n$ are representations of G,

then

$$\Gamma = c_1\Gamma_1 \oplus c_2\Gamma_2 \oplus \cdots \oplus c_n\Gamma_n = \sum_i c_i\Gamma_i \tag{7.28}$$

is also a representation of G, where here the c_i symbolically represents the number of times Γ_i occurs (appears) in Γ. Thus, in our previous example, if $\Gamma_1 = \Gamma_2$

$$\Gamma_3 = \Gamma_1 \oplus \Gamma_1 = 2\Gamma_1 \tag{7.29}$$

The special addition sign $\oplus$ is used to remind us in what sense we are using the term addition. We shall often simply use "+" but shall infer this special kind of addition whenever representations are involved.

Problem

7.10 Form a two-dimensional representation by adding the one-dimensional representations

$$\Gamma_1: \begin{matrix} e \xrightarrow{f} 1 \\ a \longrightarrow 1 \end{matrix}, \qquad \Gamma_2: \begin{matrix} e \xrightarrow{f} 1 \\ a \longrightarrow -1 \end{matrix}$$

of the group of order 2.

The addition of representations will be important later when we discuss irreducible representations, because basically, an irreducible representation is one that cannot be reduced to a form that shows it to be a sum of representations of lower dimensionality. It should be clear, for example, that any one-dimensional representation is irreducible. Given an arbitrary representation we will usually want to find just what sum of irreducible representations it consists of, another important reason for understanding the addition of representations.

DIRECT-PRODUCT REPRESENTATIONS

Just as we "sum" representations to obtain additional representations so we can "multiply" representations to obtain other representations. This is done in the sense of the direct product (Kronecker product) of matrices and can lead to either new representations of the same group or a representation of a higher order group. We shall discuss both of these in turn, but, since both involve a knowledge of the concept of the direct product of matrices, let us briefly review this idea.

The Kronecker or direct product of two square matrices **A**, **B** of dimensions d_A and d_B is another square matrix $\mathbf{C} = \mathbf{A} \otimes \mathbf{B}$ where the dimension of **C**, d_C, is the product of the dimensions of **A** and **B**

$$\mathbf{C} = \mathbf{A} \otimes \mathbf{B} \qquad d_C = d_A d_B$$
$$(C)_{ij,kl} = A_{ik}B_{jl} \tag{7.30}$$

The matrix elements of **C** are given four indices instead of the usual two in order to more readily relate **C** to **A** and **B**. To illustrate the formation of a direct product we write out the example below for two order 2 matrices

$$\begin{pmatrix} A_{11} & A_{12} \\ A_{21} & A_{22} \end{pmatrix} \otimes \begin{pmatrix} B_{11} & B_{12} \\ B_{21} & B_{22} \end{pmatrix} = \left(\begin{array}{cc:cc} A_{11}B_{11} & A_{11}B_{12} & A_{12}B_{11} & A_{12}B_{12} \\ A_{11}B_{21} & A_{11}B_{22} & A_{12}B_{21} & A_{12}B_{22} \\ \hdashline A_{21}B_{11} & A_{21}B_{12} & A_{22}B_{11} & A_{22}B_{12} \\ A_{21}B_{21} & A_{21}B_{22} & A_{22}B_{21} & A_{22}B_{22} \end{array}\right) \tag{7.31}$$

Here the dashed lines are added to call attention to the substructure of the direct product. (The direct product may also be defined for nonsquare matrices but we shall not consider it.) When multiplying direct-product matrices we find

$$(CD)_{ij,kl} = \sum_{n,m} C_{ij,nm} D_{nm,kl} \tag{7.32}$$

Problems

7.11 Show that

$$(\mathbf{A} \otimes \mathbf{B})(\mathbf{C} \otimes \mathbf{D}) = (\mathbf{AC}) \otimes (\mathbf{BD})$$

that is, the matrix product of two direct products is the direct product of the two matrix products.

7.12 Show that the direct product of two diagonal matrices is also a diagonal matrix and that the direct product of two unit matrices is also a unit matrix.

The above definitions and exercises show us that if

$$\begin{aligned} \mathbf{A}_1\mathbf{A}_2 &= \mathbf{A}_3 \\ \mathbf{B}_1\mathbf{B}_2 &= \mathbf{B}_3 \end{aligned} \tag{7.33}$$

then

$$\begin{aligned} (\mathbf{A}_1 \otimes \mathbf{B}_1)(\mathbf{A}_2 \otimes \mathbf{B}_2) &= (\mathbf{A}_1\mathbf{A}_2) \otimes (\mathbf{B}_1\mathbf{B}_2) \\ &= \mathbf{A}_3 \otimes \mathbf{B}_3 \end{aligned} \tag{7.34}$$

Consider first the case where $\Gamma_1 = \{\mathbf{D}^{(1)}(s)\}$ and $\Gamma_2 = \{\mathbf{D}^{(2)}(s)\}$ are two representations of the *same* group G. Then symbolically $\Gamma_3 = \Gamma_1 \otimes \Gamma_2 = \{\mathbf{D}^{(1)}(s) \otimes \mathbf{D}^{(2)}(s)\}$ will also be a representation of G. Thus, since

$$\begin{aligned} \mathbf{D}^{(1)}(s)\mathbf{D}^{(1)}(t) &= \mathbf{D}^{(1)}(st) \\ \mathbf{D}^{(2)}(s)\mathbf{D}^{(2)}(t) &= \mathbf{D}^{(2)}(st) \end{aligned} \tag{7.35}$$

the above equation shows us that

$$(\mathbf{D}^{(1)}(s) \otimes \mathbf{D}^{(2)}(s))(\mathbf{D}^{(1)}(t) \otimes \mathbf{D}^{(2)}(t)) = \mathbf{D}^{(1)}(st) \otimes \mathbf{D}^{(2)}(st) \tag{7.36}$$

This type of direct product is called a direct product *within* a group since we form a representation of G from two other representations of the *same* group G. Clearly if d_1 and d_2 are the dimensions of Γ_1 and Γ_2, $\Gamma_1 \otimes \Gamma_2$ is of dimension $d_1 d_2$.

We may see how the direct product within a group arises with perhaps greater insight by considering the basis sets for Γ_1 and Γ_2. For example, if $\hat{s}$ is the operator associated with $s \in G$, then for the Γ_1 basis $\{\varphi_i^{(1)}(i = 1 \text{ to } d_1)\}$

$$\hat{s}\varphi_i^{(1)} = \sum_j \varphi_j^{(1)} D^{(1)}(s)_{ji} \tag{7.37}$$

with a similar equation for the Γ_2 basis set $\{\chi_j^{(2)}(j = 1 \text{ to } d_2)\}$. *Define* now the operation of $\hat{s}$ on the *product* basis $\{\varphi_i^{(1)}\chi_j^{(2)}\}$ as

$$\hat{s}(\varphi_i^{(1)}\chi_j^{(2)}) \equiv (\hat{s}\varphi_i^{(1)})(\hat{s}\chi_j^{(2)}) \tag{7.38}$$

so that

$$\begin{aligned}
\hat{s}(\varphi_i^{(1)}\chi_j^{(2)}) &= \sum_{k,l} \varphi_k^{(1)} D^{(1)}(s)_{ki} \chi_l^{(2)} D^{(2)}(s)_{lj} \\
&= \sum_{kl} \varphi_k^{(1)} \chi_l^{(2)} D^{(1)}(s)_{ki} D^{(2)}(s)_{lj} \\
&= \sum_{kl} (\varphi_k^{(1)}\chi_l^{(2)})(\mathbf{D}^{(1)}(s) \times \mathbf{D}^{(2)}(s))_{kl,ij}
\end{aligned} \tag{7.39}$$

and the $d_1 d_2$ functions $\{\varphi_i^{(1)}\chi_j^{(2)}\}$ form a basis for a $d_1 d_2$-dimensional representation of G, $\Gamma = \{\mathbf{D}^{(1)}(s) \otimes \mathbf{D}^{(2)}(s)\}$. The action of $\hat{s}$ operating simultaneously on $\varphi_i^{(1)}$ and $\chi_j^{(2)}$ may be visualized by the example of a rotation; in a "single" basis $\hat{s}$ means "rotate a vector," whereas in a product basis the operator equivalent of the element s means "rotate vectors in the two spaces simultaneously." Clearly, the process can be extended to products of as many basis sets as we like; that is, if $\Gamma_1, \Gamma_2, \ldots, \Gamma_n$ are representations of G, then symbolically (knowing now the significance of $\otimes$ we will often use simply "$\times$")

$$\Gamma = c_1\Gamma_1 \times c_2\Gamma_2 \times \cdots \times c_n\Gamma_n \tag{7.40}$$

is also a representation of G. The c_i's remind us that Γ_i can occur as many times as we like in the direct product (or, if $c_i = 0$, not at all).

This particular type of direct product is sufficiently important that we will take a few more lines to point out how it appears in our later applications to underscore its value. The great value of group theory in quantum mechanics lies in being able to specify how quantum mechanical functions transform under the operations of the group involved. Knowledge of the transformation properties often allow us to quickly determine if certain integrals must be zero "by symmetry." Thus, if we know the representations that go along with the two sets of functions $\{\varphi_i^{(1)}\}$ and $\{\chi_j^{(2)}\}$

(the similarity to our earlier notation is intentional) we know the representation (transformational properties) of a product like $\varphi_i^{(1)}\chi_j^{(2)}$; such a product may occur in a integral where φ and χ refer to the same set of particles or it may arise as a wave function for a product system composed of the particles described by the φ-set *plus* those described by the χ-set. Our applications later on will dwell extensively on this aspect of direct products within a group.

Problem

7.13 Construct the direct-product representation of C_2, $\Gamma = \Gamma_R \times \Gamma_R$, where Γ_R is the regular representation of C_2.

The other type of direct-product representation has to do with forming a representation for a direct-product group. Recall that a group G is the direct product of its subgroups H, K if H and K are both normal subgroups and have only e, the identity element, in common. The equivalent conditions for $G = H \times K$ were that $G = HK$ (simple product), each element $g \in G$ is given by a unique product hk, $h \in H$, $k \in K$, and, finally, that $hk = kh$ all $h \in H$, $k \in K$. If Γ_H and Γ_K are representations of H and K and $G = H \times K$, then $\Gamma_H \times \Gamma_K$ is a representation of G. Note here that, aside from the trivial cases, not only is $\Gamma_H \times \Gamma_K$ of higher dimension than both Γ_H and Γ_K but it is also a representation for a *different* group.

Problems

7.14 Given $G = H \times K$, show that $\Gamma_H \times \Gamma_K$ is a representation of G.

7.15 Given that $C_{2h} = C_2 \times C_s$ and that C_2 and C_s are isomorphic, form a four-dimensional representation of C_{2h} from the regular representations of C_2 and C_s.

7.16 Use the one-dimensional representations of C_2 (or C_s) of $\Gamma_1 = \{1, 1\}$ and $\Gamma_2 = \{1, -1\}$ to

a. form the representation $\Gamma_1 \times \Gamma_2$ of C_2

b. form a one-dimensional representation

$$\Gamma_1^{C_2} \times \Gamma_2^{C_s} \quad \text{of } C_{2h}$$

SOLUTIONS TO PROBLEMS

7.1 a. $C_2 = e, a$ where $a^2 = e$. e is clearly represented by the unit matrix; therefore, a must be given by

$$\mathbf{D}(a) = \begin{pmatrix} \sqrt{3}/2 & 1/2 \\ 1/2 & -\sqrt{3}/2 \end{pmatrix}$$

This is confirmed since its square is the unit matrix.

$$\begin{pmatrix} \sqrt{3}/2 & 1/2 \\ 1/2 & -\sqrt{3}/2 \end{pmatrix}\begin{pmatrix} \sqrt{3}/2 & 1/2 \\ 1/2 & -\sqrt{3}/2 \end{pmatrix} = \begin{pmatrix} 1 & 0 \\ 0 & 1 \end{pmatrix}$$

This is a faithful representation since all the matrices involved are distinct.

b. The mapping is given by

$$e \xrightarrow{f} \begin{pmatrix} 1 & 0 & 0 \\ 0 & 1 & 0 \\ 0 & 0 & 1 \end{pmatrix}$$

$$a \xrightarrow{f} \begin{pmatrix} 0 & 0 & 1 \\ 1 & 0 & 0 \\ 0 & 1 & 0 \end{pmatrix}$$

$$a^2 \xrightarrow{f} \begin{pmatrix} 0 & 1 & 0 \\ 0 & 0 & 1 \\ 1 & 0 & 0 \end{pmatrix}$$

where the group of order 3 is the cyclic group $\{e, a, a^2\}$. The mapping is isomorphic which means that the representative is faithful.

c. The representation is obviously not faithful since several matrices are the same. We see that it is a valid representation for it represents a mapping of the factor group C_{3v}/N where N is the invariant (normal) subgroup $N = \{E, C_3, C_3^2\}$. The multiplication table is

C_{3v}/N	N	$N\sigma_1$
N	N	$N\sigma_1$
$N\sigma_1$	$N\sigma_1$	N

where we chose $N\sigma_1$ to represent the coset of N. The multiplication table shows us that the mapping is indeed a homomorphism. Note that we are speaking of a representation of C_{3v} so that while the representation is faithful to the factor group $C_{3v}/N = C_2$ it is *not* a faithful representation of C_{3v}.

7.2 Since the order of the cyclic group $C_3 = \{e, s, s^2\}$ is 3 the mapping can be

$$s \xrightarrow{f} (e^{2\pi i/3})^m$$

where *distinct* mappings will be obtained for $m = 0$, 1, and 2. Thus, there are three mappings and three representations.

Mapping	*Representation*		
	e	s	s^2
1. $m=0,\ s \xrightarrow{f} 1$	1	1	1
2. $m=1,\ s \xrightarrow{f} e^{2\pi i/3}$	1	$e^{2\pi i/3}$	$e^{4\pi i/3}$
3. $m=2,\ s \xrightarrow{f} e^{4\pi i/3}$	1	$e^{4\pi i/3}$	$e^{8\pi i/3}$

or, if we write $\varepsilon = e^{2\pi i/3}$, $e^{4\pi i/3} = e^{-2\pi i/3} = \varepsilon^*$

1.	1	1	1
2.	1	ε	ε^*
3.	1	ε^*	ε

Note that representations (2) and (3) are complex conjugates of each other. It should not be difficult to see how simply we can write down the one-dimensional representations of C_n for any n. There are precisely n of them, and, as we shall learn, these are just the n so-called irreducible or fundamental representations of the cyclic group C_n.

7.3 $\mathbf{D}(E) = \begin{pmatrix} 1 & 0 \\ 0 & 1 \end{pmatrix}$

For σ_2, since $\hat{\sigma}_2\vec{\varepsilon}_1 = -\vec{\varepsilon}_1 - \vec{\varepsilon}_2$ and $\sigma_2\vec{\varepsilon}_2 = \vec{\varepsilon}_2$, we obtain

$$\mathbf{D}(\sigma_2)\begin{pmatrix} \alpha \\ \beta \end{pmatrix} = \begin{pmatrix} -\alpha \\ -\alpha+\beta \end{pmatrix} \Rightarrow \mathbf{D}(\sigma_2) = \begin{pmatrix} -1 & 0 \\ -1 & 1 \end{pmatrix}$$

and for C_3^2, since $\hat{C}_3^2\vec{\varepsilon}_1 = -\vec{\varepsilon}_1 - \vec{\varepsilon}_2$ and $\hat{C}_3^2\vec{\varepsilon}_2 = \vec{\varepsilon}_1$ we get

$$\mathbf{D}(C_3^2)\begin{pmatrix} \alpha \\ \beta \end{pmatrix} = \begin{pmatrix} -\alpha+\beta \\ -\alpha \end{pmatrix} \Rightarrow \mathbf{D}(C_3^2) = \begin{pmatrix} -1 & 1 \\ -1 & 0 \end{pmatrix}$$

We could, of course, have made use of the C_{3v} multiplication table to find $\mathbf{D}(\sigma_2)$ and $\mathbf{D}(C_3^2)$ since $\sigma_2 = C_3\sigma_3$ and $C_3^2 = C_3C_3$; thus, for the latter case

$$\mathbf{D}(C_3^2) = \mathbf{D}(C_3)\mathbf{D}(C_3) = \begin{pmatrix} 0 & -1 \\ 1 & -1 \end{pmatrix}\begin{pmatrix} 0 & -1 \\ 1 & -1 \end{pmatrix} = \begin{pmatrix} -1 & 1 \\ -1 & 0 \end{pmatrix}$$

7.4 $(23)(x_1x_2x_3) = (x_1x_3x_2) \Rightarrow \mathbf{D}(23) = \begin{pmatrix} 1 & 0 & 0 \\ 0 & 0 & 1 \\ 0 & 1 & 0 \end{pmatrix}$

$$(123)(x_1x_2x_3) = (x_2x_3x_1) \Rightarrow \mathbf{D}(123) = \begin{pmatrix} 0 & 0 & 1 \\ 1 & 0 & 0 \\ 0 & 1 & 0 \end{pmatrix}$$

$$(13)(x_1x_2x_3)=(x_3x_2x_1)\Rightarrow \mathbf{D}(13)=\begin{pmatrix}0&0&1\\0&1&0\\1&0&0\end{pmatrix}$$

and, of course,

$$\mathbf{D}(e)=\begin{pmatrix}1&0&0\\0&1&0\\0&0&1\end{pmatrix}$$

7.5 For S_2 the elements are e and (12), and the two-dimensional representation given by our method is

$$\mathbf{D}(e)=\begin{pmatrix}1&0\\0&1\end{pmatrix},\qquad \mathbf{D}(12)=\begin{pmatrix}0&1\\1&0\end{pmatrix}$$

For the selected elements of S_4 we readily write

$$\mathbf{D}(1423)=\begin{pmatrix}0&0&1&0\\0&0&0&1\\0&1&0&0\\1&0&0&0\end{pmatrix}\qquad \mathbf{D}(123)=\begin{pmatrix}0&0&1&0\\1&0&0&0\\0&1&0&0\\0&0&0&1\end{pmatrix}$$

$$\mathbf{D}((13)(24))=\begin{pmatrix}0&0&1&0\\0&0&0&1\\1&0&0&0\\0&1&0&0\end{pmatrix}$$

In the first case (S_2) we have a two-dimensional carrier space, while for S_4 it is four-dimensional.

7.6 The group of order 3 is unique and isomorphic to the point group C_3. Cayley's theorem states that C_3 will be isomorphic to some subgroup of S_3. We thus need to find a cyclic subgroup of S_3 generated by an element of order 3. Such a subgroup is generated by (123):

$$a=(123);\qquad a^2=(123)(123)=(132);\qquad a^3=e$$

so that the desired representation is

$$\mathbf{D}(e)=\begin{pmatrix}1&0&0\\0&1&0\\0&0&1\end{pmatrix};\qquad \mathbf{D}(a)=\begin{pmatrix}0&0&1\\1&0&0\\0&1&0\end{pmatrix};$$

$$\mathbf{D}(a^2)=\begin{pmatrix}0&1&0\\0&0&1\\1&0&0\end{pmatrix}$$

We could also have chosen (132) as the generating element; the set of matrices obtained would be exactly the same as that given above but the associations with the elements (the mapping) would be different.

7.7 For V we obtain (see problem 5.4)

V	e	a	b	c
e	e	a	b	c
a	a	e	c	b
b	b	c	e	a
c	c	b	a	e

$$e \xrightarrow{f} (1)(2)(3)(4)$$
$$a \xrightarrow{f} (12)(34)$$
$$b \xrightarrow{f} (13)(24)$$
$$c \xrightarrow{f} (14)(23)$$

and the regular representation is given by

$$\mathbf{D}^R(e) = \begin{pmatrix} 1 & 0 & 0 & 0 \\ 0 & 1 & 0 & 0 \\ 0 & 0 & 1 & 0 \\ 0 & 0 & 0 & 1 \end{pmatrix} \qquad \mathbf{D}^R(a) = \begin{pmatrix} 0 & 1 & 0 & 0 \\ 1 & 0 & 0 & 0 \\ 0 & 0 & 0 & 1 \\ 0 & 0 & 1 & 0 \end{pmatrix}$$

$$\mathbf{D}^R(b) = \begin{pmatrix} 0 & 0 & 1 & 0 \\ 0 & 0 & 0 & 1 \\ 1 & 0 & 0 & 0 \\ 0 & 1 & 0 & 0 \end{pmatrix} \qquad \mathbf{D}^R(c) = \begin{pmatrix} 0 & 0 & 0 & 1 \\ 0 & 0 & 1 & 0 \\ 0 & 1 & 0 & 0 \\ 1 & 0 & 0 & 0 \end{pmatrix}$$

7.8 Let $\mathbf{D}'(s) = \mathbf{U}^{-1}\mathbf{D}(s)\mathbf{U}$, where $\mathbf{D}(s)\mathbf{D}(t) = \mathbf{D}(st)$. Then

$$\begin{aligned} \mathbf{D}'(s)\mathbf{D}'(t) &= \mathbf{U}^{-1}\mathbf{D}(s)\mathbf{U}\mathbf{U}^{-1}\mathbf{D}(t)\mathbf{U} \\ &= \mathbf{U}^{-1}\mathbf{D}(s)\mathbf{D}(t)\mathbf{U} \\ &= \mathbf{U}^{-1}\mathbf{D}(st)\mathbf{U} \\ &= \mathbf{D}'(st) \end{aligned}$$

and $\Gamma' = \{\mathbf{U}^{-1}\mathbf{D}(s)\mathbf{U}\}$ is also a representation of G.

7.9 For Γ

$$\mathbf{D}(s)\mathbf{D}(t) = \mathbf{D}(st)$$

and by taking the complex conjugate of both sides of the equation we obtain

$$\mathbf{D}^*(s)\mathbf{D}^*(t) = \mathbf{D}^*(st)$$

which shows that Γ^* is also a representation of G. Note that $\Gamma = \Gamma^*$ only if $\mathbf{D}(s) = \mathbf{D}^*(s)$ for *every* $s \in G$. In such a case Γ is real and Γ and Γ^* are identical.

7.10 Since $\Gamma_1 = \{(1), (1)\}$ [for the elements e and $a(a^2 = e)$] and $\Gamma_2 = \{(1), (-1)\}$, we form $\Gamma_1 \oplus \Gamma_2$ where

$$\Gamma_1 \oplus \Gamma_2 = \left\{ \mathbf{D}(e) = \begin{pmatrix} 1 & 0 \\ 0 & 1 \end{pmatrix}, \mathbf{D}(a) = \begin{pmatrix} 1 & 0 \\ 0 & -1 \end{pmatrix} \right\}$$

7.11
$$\begin{aligned}
[(\mathbf{A} \times \mathbf{B})(\mathbf{C} \times \mathbf{D})]_{ij,kl} &= \sum_{n,m} (\mathbf{A} \times \mathbf{B})_{ij,nm} (\mathbf{C} \times \mathbf{D})_{nm,kl} \\
&= \sum_{n,m} (A_{in} B_{jm})(C_{nk} D_{ml}) \\
&= \sum_{n,m} (A_{in} C_{nk})(B_{jm} D_{ml}) \\
&= (\mathbf{AC})_{ik} (\mathbf{BD})_{jl} \\
&= [(\mathbf{AC}) \times (\mathbf{BD})]_{ij,kl}
\end{aligned}$$

7.12 Let **A** and **B** be diagonal

$$A_{ik} = a_i \delta_{ik} ; \qquad B_{jm} = b_j \delta_{jm}$$

Then

$$\begin{aligned}
(\mathbf{A} \times \mathbf{B})_{ij,km} &= A_{ik} B_{jm} \\
&= a_i b_j \delta_{ik} \delta_{jm}
\end{aligned}$$

so the only nonvanishing elements are those for $i = k$, $j = m$, the diagonal elements of $\mathbf{A} \times \mathbf{B}$

$$(\mathbf{A} \times \mathbf{B})_{ij,ij} = a_i b_j$$

Clearly, if **A** and **B** are *unit* matrices so that $a_i = b_j = 1$ then $\mathbf{A} \times \mathbf{B}$ is clearly also a unit matrix.

7.13 The regular representation of C_2 is readily written down:

C_2	e	a
e	e	a
a	a	e

(1)(2) $\quad \mathbf{D}(e) = \begin{pmatrix} 1 & 0 \\ 0 & 1 \end{pmatrix}$

(12) $\quad \mathbf{D}(a) = \begin{pmatrix} 0 & 1 \\ 1 & 0 \end{pmatrix}$

Thus for $\Gamma_R \times \Gamma_R$

$$\mathbf{D}^R(e) \times \mathbf{D}^R(e) = \begin{pmatrix} 1 & 0 & 0 & 0 \\ 0 & 1 & 0 & 0 \\ 0 & 0 & 1 & 0 \\ 0 & 0 & 0 & 1 \end{pmatrix} \qquad \text{(by problem 7.12)}$$

$$\mathbf{D}^R(a) \times \mathbf{D}^R(a) = \begin{pmatrix} 0 & 1 \\ 1 & 0 \end{pmatrix} \times \begin{pmatrix} 0 & 1 \\ 1 & 0 \end{pmatrix} = \begin{pmatrix} 0 & 0 & 0 & 1 \\ 0 & 0 & 1 & 0 \\ 0 & 1 & 0 & 0 \\ 1 & 0 & 0 & 0 \end{pmatrix}$$

7.14 The mappings

$$h \xrightarrow{f} \mathbf{D}^H(h), \qquad k \xrightarrow{f'} \mathbf{D}^K(k)$$

are by definition homomorphisms since Γ_H and Γ_K are representations of H and K. Given that $G = H \times K$ so that $g = hk$ (unique product) we look for a homomorphism

$$f(hk)f(h'k') = f(hkh'k') = f(hh'kk')$$

where we used the (necessary) fact that $hk = kh$. For

$$f(hk) = \mathbf{D}^H(h) \times \mathbf{D}^K(k)$$

we see that (see problem 7.11)

$$\begin{aligned}
&[\mathbf{D}^H(h) \times \mathbf{D}^K(k)][\mathbf{D}^H(h') \times \mathbf{D}^K(k')] \\
&\quad = [\mathbf{D}^H(h)\mathbf{D}^H(h')] \times [\mathbf{D}^K(k)\mathbf{D}^K(k')] \\
&\quad = [\mathbf{D}^H(hh') \times \mathbf{D}^K(kk')]
\end{aligned}$$

and the homomorphism condition is satisfied.

7.15 Since C_2 and C_s are isomorphic, their regular representations are identical and given by

$$\Gamma_R = \left\{ \begin{pmatrix} 1 & 0 \\ 0 & 1 \end{pmatrix}, \begin{pmatrix} 0 & 1 \\ 1 & 0 \end{pmatrix} \right\}$$

Thus, given

$$\mathbf{D}(e(C_2)) = \begin{pmatrix} 1 & 0 \\ 0 & 1 \end{pmatrix} \qquad \mathbf{D}(e(C_s)) = \begin{pmatrix} 1 & 0 \\ 0 & 1 \end{pmatrix}$$

$$\mathbf{D}(C_2) = \begin{pmatrix} 0 & 1 \\ 1 & 0 \end{pmatrix} \qquad \mathbf{D}(\sigma_h) = \begin{pmatrix} 0 & 1 \\ 1 & 0 \end{pmatrix}$$

we obtain for

$$\mathbf{D}(ee) = \mathbf{D}(e) = \begin{pmatrix} 1 & 0 \\ 0 & 1 \end{pmatrix} \times \begin{pmatrix} 1 & 0 \\ 0 & 1 \end{pmatrix} = \begin{pmatrix} 1 & 0 & 0 & 0 \\ 0 & 1 & 0 & 0 \\ 0 & 0 & 1 & 0 \\ 0 & 0 & 0 & 1 \end{pmatrix}$$

$$\mathbf{D}(e\sigma_h) = \mathbf{D}(\sigma_h) = \begin{pmatrix} 1 & 0 \\ 0 & 1 \end{pmatrix} \times \begin{pmatrix} 0 & 1 \\ 1 & 0 \end{pmatrix} = \begin{pmatrix} 0 & 1 & 0 & 0 \\ 1 & 0 & 0 & 0 \\ 0 & 0 & 0 & 1 \\ 0 & 0 & 1 & 0 \end{pmatrix}$$

$$\mathbf{D}(C_2e) = \mathbf{D}(C_2) = \begin{pmatrix} 0 & 1 \\ 1 & 0 \end{pmatrix} \times \begin{pmatrix} 1 & 0 \\ 0 & 1 \end{pmatrix} = \begin{pmatrix} 0 & 0 & 1 & 0 \\ 0 & 0 & 0 & 1 \\ 1 & 0 & 0 & 0 \\ 0 & 1 & 0 & 0 \end{pmatrix}$$

$$\mathbf{D}(C_2\sigma_h)=\mathbf{D}(i)=\begin{pmatrix}0 & 1\\ 1 & 0\end{pmatrix}\times\begin{pmatrix}0 & 1\\ 1 & 0\end{pmatrix}=\begin{pmatrix}0 & 0 & 0 & 1\\ 0 & 0 & 1 & 0\\ 0 & 1 & 0 & 0\\ 1 & 0 & 0 & 0\end{pmatrix}$$

7.16 Since we are dealing with one-dimensional matrices, matrix multiplication is simple.

(a) $\Gamma_1\times\Gamma_2=\{(1)(1),(1)(-1)\}=\{1,-1\}=\Gamma_2$
Note that $\Gamma_1\times\Gamma_1=\Gamma_1=\Gamma_2\times\Gamma_2$

(b) $\Gamma_1\times\Gamma_2=\{(1)(1)=\mathbf{D}(ee),$
$(1)(-1)=\mathbf{D}(e\sigma_h),$
$(1)(1)=\mathbf{D}(C_2e),$
$(1)(-1)=\mathbf{D}(C_2\sigma_h)\}$

Note that $\Gamma_1\times\Gamma_2$ is not a faithful representation.

8
Irreducible Representations

We spoke earlier of the fact that although we may find essentially an infinite number of representations (not counting equivalent representations) of a group, there exists for each group a *finite* number of basic or fundamental representations which are called the *irreducible representations* of the group. Knowledge of the properties of these irreducible representations is of paramount importance in the application of group theory to quantum mechanical systems. Since we normally begin with a reducible (nonirreducible) representation, we must also know how to decompose (reduce) such a representation into its irreducible components. In this chapter we deal with these most important questions.

CRITERIA FOR IRREDUCIBILITY

In Chapter 7 we noted that we could "add" representations to obtain additional representations of a group. Thus, $\Gamma = \Gamma_1 + \Gamma_2 = \{\mathbf{D}(s)\}$ is a representation of G where

$$\mathbf{D}(s) = \begin{pmatrix} \mathbf{D}^{(1)}(s) & 0 \\ 0 & \mathbf{D}^{(2)}(s) \end{pmatrix} \tag{8.1}$$

More generally,

$$\mathbf{D}(s) = \begin{pmatrix} \mathbf{D}^{(1)}(s) & & & & \mathbf{0} \\ & \mathbf{D}^{(2)}(s) & & & \\ & & \cdot & & \\ & & & \cdot & \\ \mathbf{0} & & & & \mathbf{D}^{(n)}(s) \end{pmatrix} \tag{8.2}$$

and also represents G if $\Gamma_1, \Gamma_2, \ldots, \Gamma_n$ do. If *all* the matrices $\mathbf{D}(s)$ of a representation Γ are of the above blocked form, Γ is said to be a *reducible representation.* Note that it is necessary for *all* the matrices $\mathbf{D}(s)$ to block in exactly the same way. Note, too, that while $\mathbf{D}(s)$ may not be blocked initially, if it is equivalent to a blocked set—that is, if it can undergo a similarity transformation to blocked form—it is still said to be a reducible representation. Thus, the two representations of C_2

$$\Gamma_1: \quad \mathbf{D}(e) = \begin{pmatrix} 1 & 0 \\ 0 & 1 \end{pmatrix}, \quad \mathbf{D}(a) = \begin{pmatrix} 1 & 0 \\ 0 & -1 \end{pmatrix}; \qquad \Gamma_2: \quad \mathbf{D}(e) = \begin{pmatrix} 1 & 0 \\ 0 & 1 \end{pmatrix}, \quad \mathbf{D}(a) = \begin{pmatrix} \sqrt{3}/2 & 1/2 \\ 1/2 & -\sqrt{3}/2 \end{pmatrix}$$

are related by a similarity transformation and are accordingly *both* reducible. It is clear that the form of a representation may not immediately reveal its reducible or irreducible nature.

Given that $\mathbf{D}(s)$ is of the above blocked form (or is equivalent to a blocked form), for all $s \in G$ one must then investigate separately the sets $\{\mathbf{D}^{(i)}(s)\}$ to test their reducibility. A matrix set $\{\mathbf{D}^{(\alpha)}(s)\}$, which can be reduced no further, is called an *irreducible representation.* We mentioned before that all one-dimensional representations are obviously irreducible.

Actually, the criteria for reducibility is less stringent than stated above. To illustrate this point consider a representation where each matrix has the form

$$\mathbf{D}(s) = \begin{pmatrix} \mathbf{D}^{(1)}(s) & \mathbf{A}(s) \\ 0 & \mathbf{D}^{(2)}(s) \end{pmatrix} \tag{8.3}$$

If $\mathbf{D}^{(1)}$ is an $n \times n$ matrix and $\mathbf{D}^{(2)}$ $m \times m$, then $\mathbf{D}$ is $(n+m)\times(n+m)$ and $\mathbf{A}$ is a rectangular matrix of m columns and n rows. It is readily shown in terms of these submatrices that

$$\begin{pmatrix} \mathbf{D}^{(1)}(s) & \mathbf{A}(s) \\ 0 & \mathbf{D}^{(2)}(s) \end{pmatrix} \begin{pmatrix} \mathbf{D}^{(1)}(t) & \mathbf{A}(t) \\ 0 & \mathbf{D}^{(2)}(t) \end{pmatrix}$$

$$= \begin{pmatrix} \mathbf{D}^{(1)}(s)\mathbf{D}^{(1)}(t) & \mathbf{D}^{(1)}(s)\mathbf{A}(t) + \mathbf{A}(s)\mathbf{D}^{(2)}(t) \\ 0 & \mathbf{D}^{(2)}(s)\mathbf{D}^{(2)}(t) \end{pmatrix} \tag{8.4}$$

Since the larger dimension set $\{\mathbf{D}(s)\}$ is a representation it follows that

$$\begin{aligned} \mathbf{D}^{(1)}(s)\mathbf{D}^{(1)}(t) &= \mathbf{D}^{(1)}(st) \\ \mathbf{D}^{(2)}(s)\mathbf{D}^{(2)}(t) &= \mathbf{D}^{(2)}(st) \\ \mathbf{D}^{(1)}(s)\mathbf{A}(t) + \mathbf{A}(s)\mathbf{D}^{(2)}(t) &= \mathbf{A}(st) \end{aligned} \tag{8.5}$$

where the first two equations show us that $\{\mathbf{D}^{(1)}(s)\}$ and $\{\mathbf{D}^{(2)}(s)\}$ are also representations of G. The matrix form above is an example of a form said to be *superdiagonal.* It can be shown that such a superdiagonal form can always be brought to diagonal form by a suitable similarity transformation so that once a representation is recognized to be in superdiagonal form we know it is reducible.

Still another way of recognizing (or, perhaps, stating) the reducibility criterion is to note whether or not the carrier space of a particular representation has any stable subspaces. For example, if

$$\mathbf{D}(s) = \begin{pmatrix} \mathbf{D}^{(1)}(s) & 0 \\ 0 & \mathbf{D}^{(2)}(s) \end{pmatrix} \begin{matrix} \updownarrow n \\ \updownarrow m \end{matrix} \quad \text{(columns: } \leftarrow n \rightarrow \;\; \leftarrow m \rightarrow \text{)} \tag{8.6}$$

we can see that vectors of the form

$$\begin{pmatrix} x_1 \\ x_2 \\ \cdot \\ \cdot \\ \cdot \\ x_n \\ 0 \\ 0 \\ \cdot \\ \cdot \\ \cdot \\ 0 \end{pmatrix} \begin{matrix} \left.\vphantom{\begin{matrix} x \\ x \\ x \\ x \\ x \\ x \end{matrix}}\right\} \text{first } n \text{ components} \left\{\vphantom{\begin{matrix} x \\ x \\ x \\ x \\ x \\ x \end{matrix}}\right. \\ \left.\vphantom{\begin{matrix} x \\ x \\ x \\ x \\ x \\ x \end{matrix}}\right\} \text{last } m \text{ components} \left\{\vphantom{\begin{matrix} x \\ x \\ x \\ x \\ x \\ x \end{matrix}}\right. \end{matrix} \begin{pmatrix} 0 \\ 0 \\ \cdot \\ \cdot \\ \cdot \\ 0 \\ x_{n+1} \\ x_{n+2} \\ \cdot \\ \cdot \\ \cdot \\ x_{n+m} \end{pmatrix}$$

when operated on by the $\mathbf{D}(s)$ yield vectors restricted to the same subspace. Thus, $\{\phi_i, i = 1 \text{ to } n\}$ forms a carrier space for an n-dimensional representation of the group in question and likewise the set $\{\phi_i, i = n+1 \text{ to } n+m\}$ spans an m-dimensional carrier space for an m-dimensional representation. Often stable subspaces are readily recognizable. For example, recall the example treated previously of the group C_{3v} where in order to form a representation we choose a carrier space consisting of any two noncolinear vectors in a plane normal to the threefold rotation axis. We could easily have formed a three-dimensional representation by adding a third vector out of the plane normal to C_3. We would then see that the previous two-dimensional subspace is a stable subspace of the three-dimensional space in that all the operations of C_{3v} do not take a vector originally in the subspace out of it. Accordingly, we would correctly assess the three-dimensional representation as reducible.

It is perhaps clear at this point that it is generally easier to recognize reducibility than it is to actually perform the reduction to irreducible form. Fortunately, what needs to be done in practice is to find the composition of the representation in terms of the irreducible representations that sum to give it, and this is fairly readily done as we shall see later. We need to know the properties of irreducible representations and those have been conveniently tabulated for us in tables.

SOME BASIC THEOREMS

The orthogonality relations to be presented in this next section contain in concise form some very important properties of irreducible representations. We shall state without proof several central theorems that are necessary to establish those relations. The proofs involved are not overly complicated but are a bit tedious; the interested reader may find them in the more advanced books on group theory. The important theorems and some brief commentary follow.

1. *Any representation of a finite group is equivalent to a unitary representation.* This theorem can be extended to certain infinite groups; the problem in generalizing it to all groups has to do with the infinite number of elements in an infinite group. It is, however, perfectly general for finite groups and indicates that any nonunitary representation can be transformed into a unitary one by a suitable similarity transformation. Since such can always be done, we can always assume our representations (for finite groups) to be unitary. This is convenient since it allows us to readily write down $\mathbf{D}(s^{-1})$ in terms of $\mathbf{D}(s)$. Thus, since $\mathbf{D}(s^{-1}) = \mathbf{D}(s)^{-1}$ by the homomorphism condition, if Γ is unitary $\mathbf{D}(s)^{-1} = \mathbf{D}(s)^{\dagger}$ so that $\mathbf{D}(s^{-1}) = \mathbf{D}(s)^{\dagger}$.
2. *The only type of matrix that will commute with all the matrices of an irreducible representation is a constant matrix. If a nonconstant matrix exists that will commute with all the matrices of a representation, the representation is reducible.* These statements compose part of the famous Schur's lemmas (Schur's first lemma) and give not only a concise criterion (in principle) for reducibility but also provide the basis for constructing the very important orthogonality relations. By a constant matrix we mean some multiple of the unit matrix.
3. *If a matrix* $\mathbf{M}$ *exists such that* $\mathbf{D}^{(\alpha)}(s)\mathbf{M} = \mathbf{M}\mathbf{D}^{(\beta)}(s)$, all $s \in G$, *for two irreducible representations* Γ_α *and* Γ_β, *then either* $\mathbf{M} = 0$ *or, if* $\mathbf{M} \neq 0$, Γ_α *and* Γ_β *are equivalent irreducible representations.* This completes the statement of Schur's lemmas (Schur's second lemma) and provides us with the basis for establishing the orthogonality relations. If Γ_α is of dimension d_α and Γ_β of dimension d_β, clearly $\mathbf{M}$ is a matrix of d_α rows and d_β columns. Obviously if $d_\alpha \neq d_\beta$ $\mathbf{M} = 0$ by the theorem since there is no way representations (irreducible or not) of different dimension can be equivalent. $\mathbf{M} = 0$ also obtains if $d_\alpha = d_\beta$ but $\Gamma_\alpha \neq \Gamma_\beta$; if the representations are equivalent, then $\mathbf{D}^{(\alpha)}(s) = \mathbf{M}\mathbf{D}^{(\beta)}(s)\mathbf{M}^{-1}$ and $\mathbf{M}$ is simply the similarity transformation matrix relating the two irreducible matrices.

Problem

8.1 Let $\{\phi_i\}$ be a basis for a representation $\Gamma = \{\mathbf{D}(s)\}$ of G of dimension d and let $\{\chi_j = \sum_i \phi_i C_{ij}\}$ be a basis of a stable subspace for a representation $\Gamma' = \{\mathbf{D}'(s)\}$ of lower dimensionality. Show that this leads to the relation

$$\mathbf{D}(s)\mathbf{C} = \mathbf{C}\mathbf{D}'(s)$$

where $(\mathbf{C})_{ij} = C_{ij}$, supporting the theorem which predicts that Γ is not an irreducible representation.

THE ORTHOGONALITY RELATIONS

Using Schur's lemmas we proceed to derive the important orthogonality relations. We begin by observing that

$$\mathbf{M} = \sum_{s \in G} \mathbf{D}^{(\alpha)}(s)\mathbf{A}\mathbf{D}^{(\beta)}(s^{-1}) \tag{8.7}$$

satisfies the equations

$$\mathbf{D}^{(\alpha)}(t)\mathbf{M} = \mathbf{M}\mathbf{D}^{(\beta)}(t), \qquad \text{all } t \in G \tag{8.8}$$

where Γ_α, Γ_β are two irreducible representations of G and $\mathbf{A}$ is an as yet unspecified matrix (d_α rows, d_β columns).

Problems

8.2 Show that $\mathbf{M}$ does indeed satisfy the above equation. Remember the rearrangement theorem.

8.3 Show that, more generally,

$$\mathbf{B} = \sum_{s \in g} \mathbf{D}^{(\alpha)}(s)\mathbf{A}\mathbf{D}^{(\beta)}(as^{-1}), \text{ any } a \in G$$

also satisfies the above equation.

Now, since Γ_α and Γ_β are irreducible representation of G Schur's lemma (theorem 3) tell us that unless Γ_α and Γ_β are equivalent, $\mathbf{M} = 0$ identically. That is,

$$M_{il} = \sum_{\substack{s \in G, \\ p,q}} D^{(\alpha)}(s)_{ip} A_{pq} D^{(\beta)}(s^{-1})_{ql} = 0 \quad (\Gamma_\alpha \neq \Gamma_\beta) \tag{8.9}$$

for *arbitrary* $\mathbf{A}$. Since we are interested specifically in relations between the irreducible representations, we can arbitrarily select $\mathbf{A}$ to "drop out" of the above equation; thus, let

$$A_{pq} = \delta_{pj}\delta_{qk} \tag{8.10}$$

that is, let $\mathbf{A}$ consist of only the one nonzero element $A_{jk} = \lambda$ so that our relation becomes

$$\sum_{s \in G} D^{(\alpha)}(s)_{ij} D^{(\beta)}(s^{-1})_{kl} = 0 \quad (\Gamma_\alpha \neq \Gamma_\beta) \tag{8.11}$$

What happens if Γ_α and Γ_β are equivalent? According to our earlier discussion **M** *can* be nonzero, and if it is, it must be the similarity transformation that relates the two equivalent representations so that $\mathbf{D}^{(\alpha)}(s) = \mathbf{M}\mathbf{D}^{(\beta)}(s)\mathbf{M}^{-1}$. We shall only consider those instances where either Γ_α and Γ_β are different representations (nonequivalent) or are identical representations. If $\Gamma_\alpha = \Gamma_\beta$ Schur's lemma (theorem 2) indicates that

$$\mathbf{M} = c\mathbf{1} \tag{8.12}$$

where c is some constant and **1** is the unit matrix of dimension $d_\alpha = d_\beta$. Since **A** is still arbitrary and now

$$\sum_{\substack{s \in G,\\ p,q}} D^{(\alpha)}(s)_{ip} A_{pq} D^{(\alpha)}(s^{-1})_{ql} = c\delta_{il} \tag{8.13}$$

by selecting $A_{pq} = \delta_{pk}\delta_{jq}$ we obtain

$$\sum_{s \in G} D^{(\alpha)}(s)_{ik} D^{(\alpha)}(s^{-1})_{jl} = c\delta_{il} \tag{8.14}$$

To determine what the constant c must be we investigate the sum of diagonal elements; that is, set $i = l$ and sum over i

$$\sum_i \sum_{s \in G} D^{(\alpha)}(s)_{ik} D^{(\alpha)}(s^{-1})_{ji} = c\sum_i \delta_{ii}$$

$$\sum_i \sum_{s \in G} D^{(\alpha)}(s^{-1})_{ji} D^{(\alpha)}(s)_{ik} = cd_\alpha \tag{8.15}$$

$$\sum_{s \in G} D^{(\alpha)}(e)_{jk} = cd_\alpha$$

$\mathbf{D}^{(\alpha)}(e)$ is diagonal, of course, and since the summand is independent of s, the $\sum_{s \in G}$ gives us just $°G$, so that

$$cd_\alpha = \sum_{s \in G} \delta_{jk} = °G\delta_{jk}$$

$$c = \frac{°G}{d_\alpha}\delta_{jk} \tag{8.16}$$

Combining this result with the earlier one for different representations we can state the *master orthogonality relations:* If Γ_α and Γ_β are nonequivalent irreducible representations of G then

$$\sum_{s \in G} D^{(\alpha)}(s)_{ik} D^{(\beta)}(s^{-1})_{jl} = \frac{°G}{d_\alpha}\delta_{\alpha\beta}\delta_{il}\delta_{kj} \tag{8.17}$$

If, in addition, Γ_α and Γ_β are unitary

$$\sum_{s \in g} D^{(\alpha)}(s)_{ik} D^{(\beta)}(s)^*_{lj} = \frac{°G}{d_\alpha}\delta_{\alpha\beta}\delta_{il}\delta_{kj} \tag{8.18}$$

The above are rather formidable expressions but very powerful ones. From them spring the results that are so useful in physical applications.

Problem

8.4 The three (and only three) irreducible representations of C_{3v} consist of two one-dimensional and one two-dimensional representation and may be written as

	E	σ_1	σ_2	σ_3	C_3	C_3^2
Γ_1	1	1	1	1	1	1
Γ_2	1	-1	-1	-1	1	1
Γ_3	$\begin{pmatrix}1 & 0\\ 0 & 1\end{pmatrix}$	$\begin{pmatrix}-1 & 0\\ 0 & 1\end{pmatrix}$	$\begin{pmatrix}1/2 & -\sqrt{3}/2\\ -\sqrt{3}/2 & -1/2\end{pmatrix}$	$\begin{pmatrix}1/2 & \sqrt{3}/2\\ \sqrt{3}/2 & -1/2\end{pmatrix}$	$\begin{pmatrix}-1/2 & \sqrt{3}/2\\ -\sqrt{3}/2 & -1/2\end{pmatrix}$	$\begin{pmatrix}-1/2 & -\sqrt{3}/2\\ \sqrt{3}/2 & -1/2\end{pmatrix}$

Verify the master orthogonality relations noting that Γ_1, Γ_2 and Γ_3 above are unitary representations.

The form of the orthogonality relations for unitary representations is suggestive of the scalar product of vectors. Indeed, if we define a vector specified by the three indices α, i, j in an "element" space (${}^\circ G$-dimensional) with the sth component given by

$$\sqrt{\frac{d_\alpha}{{}^\circ G}}\, D^{(\alpha)}(s)_{ij}$$

we see that we obtain a set of orthonormal vectors in a complex space, the vector orthogonality relations being simply our master orthogonality relation. For each representation Γ_α there are d_α^2 such vectors, or, for all representations a total of $\sum_\alpha d_\alpha^2$ orthogonal vectors in an ${}^\circ G$-dimensional space. Now a general property of proper vector spaces is that one can have *no more* linearly independent orthogonal vectors than the dimensionality of the space; think of the readily visualized cases of real two- and three-dimensional spaces in the laboratory. Accordingly,

$$\sum_\alpha d_\alpha^2 \leq {}^\circ G \tag{8.19}$$

and the master orthogonality relations interpreted in the above manner have led us to a condition on the allowed dimensionalities of irreducible representations. Since ${}^\circ G$ is finite for finite groups, it is clear that the number of irreducible representations must be finite. In the next chapter we shall see that the above inequality is actually an equality.

The use of the $D^{(\alpha)}(s)_{ij}$ as vector components is also very useful in the construction of certain operators. We shall have more to say concerning this topic later and shall only illustrate the idea with an exercise.

Problem

8.5 Let the *sets* $\{\phi_i^{(\alpha)}\}$, $\alpha = 1, 2, \ldots$ comprise sets of basis functions for the various *irreducible* representations of G. Show that the operator

$$\hat{P}_{kl}^{(\alpha)}=\frac{d_\alpha}{{}^oG}\sum_{s\in G} D^{(\alpha)}(s)^*_{kl}\hat{O}_s$$

has the properties that

$$\hat{P}_{kl}^{(\alpha)}\phi_j^{(\beta)}=0,\qquad \alpha\neq\beta$$
$$\hat{P}_{kl}^{(\alpha)}\phi_j^{(\alpha)}=0,\qquad j\neq l$$
$$\hat{P}_{kl}^{(\alpha)}\phi_l^{(\alpha)}=\phi_k^{(\alpha)}$$

SOLUTIONS TO PROBLEMS

8.1 The idea here is to relate Schur's lemmas and the idea of stable subspaces. The conditions of the problem give us

$$s\rightarrow\hat{O}_s$$
$$\hat{O}_s\phi_i=\sum_j \phi_j D_{ji}(s)$$
$$\hat{O}_s\chi_l=\sum_p \chi_p D'_{pl}(s)$$

Thus, since

$$\chi_j=\sum_i \phi_i C_{ij}$$
$$\hat{O}_s\chi_j=\sum_i \hat{O}_s\phi_i C_{ij}=\sum_{il}\phi_l D_{li}(s)C_{ij}$$

and

$$\hat{O}_s\chi_j=\sum_p \chi_p D'_{pj}(s)=\sum_{pq}\phi_q C_{qp}D'(s)_{pj}$$

so that

$$\sum_{pq}\phi_q C_{qp}D'(s)_{pj}=\sum_{il}\phi_l D(s)_{li}C_{ij}$$

Since the $\{\phi_i\}$ must be linearly independent functions it follows that

$$\sum_i C_{ki}D'(s)_{ij}=\sum_i D_{ki}(s)C_{ij}$$
$$\mathbf{CD'}(s)=\mathbf{D}(s)\mathbf{C}$$

C is not a null matrix so the presence of a stable subspace leads to this particular way of expressing the reducibility of Γ.

8.2

$$\mathbf{D}^{(\alpha)}(t)\mathbf{M}=\sum_{s\in G}\mathbf{D}^{(\alpha)}(t)\mathbf{D}^{(\alpha)}(s)\mathbf{A}\mathbf{D}^{(\beta)}(s^{-1})$$
$$=\sum_s \mathbf{D}^{(\alpha)}(ts)\mathbf{A}\mathbf{D}^{(\beta)}(s^{-1})$$

Multiplying on the right by the unit matrix in the form of

$$\mathbf{1}=\mathbf{D}^{(\beta)}(e)=\mathbf{D}^{(\beta)}(t^{-1}t)=\mathbf{D}^{(\beta)}(t^{-1})\mathbf{D}^{(\beta)}(t)$$

and realizing by the rearrangement theorem that

$$\sum_{s\in G} f(s) = \sum_{s\in G} f(ts)$$

that is, the two sums still cover all elements in the group, we get

$$\begin{aligned}\mathbf{D}^{(\alpha)}(t)\mathbf{M} &= \sum_s \mathbf{D}^{(\alpha)}(ts)\mathbf{A}\mathbf{D}^{(\beta)}(s^{-1})\mathbf{D}^{(\beta)}(t^{-1})\mathbf{D}^{(\beta)}(t)\\ &= \sum_s \mathbf{D}^{(\alpha)}(ts)\mathbf{A}\mathbf{D}^{(\beta)}((ts)^{-1})\mathbf{D}^{(\beta)}(t)\\ &= \mathbf{M}\mathbf{D}^{(\beta)}(t)\end{aligned}$$

8.3 This is simply done by the same method given in problem 8.2 and shows us that we can form quite a number of matrices **M** satisfying

$$\mathbf{D}^{(\alpha)}(s)\mathbf{M} = \mathbf{M}\mathbf{D}^{(\beta)}(s)$$

8.4 Solution stated in problem.

8.5 Consider the action of $\hat{P}_{kl}^{(\alpha)}$ on a general basis function $\phi_j^{(\beta)}$ and employ the orthogonality relations.

$$\begin{aligned}\hat{P}_{kl}^{(\alpha)}\phi_j^{(\beta)} &= \frac{d_\alpha}{{}^oG}\sum_{s\in G} D^{(\alpha)}(s)_{kl}^*\hat{O}_s\phi_j^{(\beta)}\\ &= \frac{d_\alpha}{{}^oG}\sum_{s,p} D^{(\alpha)}(s)_{kl}^* D^{(\beta)}(s)_{pj}\phi_p^{(\beta)}\\ &= \sum_p \delta_{\alpha\beta}\delta_{kp}\delta_{lj}\phi_p^{(\beta)}\end{aligned}$$

Thus

$$\text{if } \alpha \neq \beta,\ \hat{P}_{kl}^{(\alpha)}\phi_j^{(\beta)} = 0$$
$$\text{if } \alpha = \beta,\ j \neq l,\ \hat{P}_{kl}^{(\alpha)}\phi_j^{(\alpha)} = 0$$

and

$$\text{if } \alpha = \beta,\ j = l,\ \hat{P}_{kl}^{(\alpha)}\phi_l^{(\alpha)} = \sum_p \delta_{kp}\phi_p^{(\alpha)} = \phi_k^{(\alpha)}$$

9
The Character

In this chapter we converge on the single most important item in our growing arsenal of group theoretical weapons. Essentially all the information we need in analyzing a physical system in terms of its symmetry properties is contained in the group character table. Knowledge of the properties of the character of representations will allow us to quickly determine reducibility or irreducibility, to analyze a general representation in terms of the irreducible representations it contains, and, ultimately, to provide us with the means for constructing basis sets that will serve as carrier space bases for irreducible representations.

DEFINITION AND BASIC PROPERTIES

The character χ, or trace, of a matrix is the sum of the diagonal terms of that matrix.

$$\chi = \mathrm{Tr}\,\mathbf{A} = \sum_i A_{ii} \tag{9.1}$$

where Tr ("trace") is used to indicate the action performed. The trace or character has a number of important and readily derived properties illustrated by the following exercises.

Problems

9.1 Show that the trace of a product of matrices is independent of the *cyclic* order of the matrix product; that is,

$$\mathrm{Tr}\,\mathbf{ABC} = \mathrm{Tr}\,\mathbf{BCA} = \mathrm{Tr}\,\mathbf{CAB}$$

9.2 Show that two matrices related by a similarity transformation have the same character.

9.3 Show that

a. $\mathrm{Tr}\,(\mathbf{A}\otimes\mathbf{B}) = (\mathrm{Tr}\,\mathbf{A})(\mathrm{Tr}\,\mathbf{B})$

b. $\mathrm{Tr}\,(\mathbf{A}\oplus\mathbf{B}) = (\mathrm{Tr}\,\mathbf{A}) + (\mathrm{Tr}\,\mathbf{B})$

With each matrix $\mathbf{D}(s) \in \Gamma$ can be associated its trace or character $\chi(s) = \mathrm{Tr}\,\mathbf{D}(s)$. The *character of a representation* is the *set* of characters $\chi = \{\chi(s)\}$. If Γ is of dimension d the set will contain d traces; although they may not all be distinct, all d traces are listed in χ. We indicate χ as the set of *numbers* $\{\chi(s)\}$ just as we use Γ to indicate the set of *matrices* $\{\mathbf{D}(s)\}$. Although the term "character" is used both for the trace of an individual matrix and for the set of traces associated with a particular representation, the one that is meant is usually clear from context.

As an example, consider the three irreducible representations of C_{3v} mentioned earlier:

	E	σ_1	σ_2	σ_3	C_3	C_3^2
Γ_1	1	1	1	1	1	1
Γ_2	1	−1	−1	−1	1	1
Γ_3	$\begin{pmatrix} 1 & 0 \\ 0 & 1 \end{pmatrix}$	$\begin{pmatrix} -1 & 0 \\ 0 & 1 \end{pmatrix}$	$\begin{pmatrix} 1/2 & -\sqrt{3}/2 \\ -\sqrt{3}/2 & -1/2 \end{pmatrix}$	$\begin{pmatrix} 1/2 & \sqrt{3}/2 \\ \sqrt{3}/2 & -1/2 \end{pmatrix}$	$\begin{pmatrix} -1/2 & \sqrt{3}/2 \\ -\sqrt{3}/2 & -1/2 \end{pmatrix}$	$\begin{pmatrix} -1/2 & -\sqrt{3}/2 \\ \sqrt{3}/2 & -1/2 \end{pmatrix}$

The characters of the three representations are then

	$\chi(E)$	$\chi(\sigma_1)$	$\chi(\sigma_2)$	$\chi(\sigma_3)$	$\chi(C_3)$	$\chi(C_3^2)$
$\chi^{(1)}$	1	1	1	1	1	1
$\chi^{(2)}$	1	−1	−1	−1	1	1
$\chi^{(3)}$	2	0	0	0	−1	−1

We note that the characters for one-dimensional representations are simply the matrices themselves. The following exercises illustrate several other important properties.

Problems

9.4 Show that elements belonging to the same class have the same character in any particular representation.

9.5 Show that $\chi(e) = d$, where e is the unit element and d is the dimension of the representation.

9.6 Use the previous theorem about unitary representations and finite group to show that

$$\chi(s^{-1}) = \chi(s)^*$$

for groups of finite order.

CHARACTER ORTHOGONALITY RELATIONS

The master orthogonality relations for nonequivalent irreducible representations

$$\frac{d_\alpha}{{}^\circ G} \sum_{s \in G} D^{(\alpha)}(s)_{ij} D^{(\beta)}(s^{-1})_{kn} = \delta_{\alpha\beta}\delta_{in}\delta_{jk} \tag{9.2}$$

is difficult to use in the sense that there are as many ways of writing down

Γ_α or Γ_β as there are similarity transformations. It is similar to the situation involving the choice of a unit cell in a crystal; there is no single "right" way to do it, and one must, therefore, determine a convention so that we can at least be consistent. The same is true with irreducible representations. Fortunately, we can derive orthogonality relations that depend on the characters and since the characters are invariant under a similarity transformation we end up with an "absolute" set of orthogonality relations. We do pay a price for this simplification, and the resulting orthogonality relations are less powerful than the master set; however, the application of the theory becomes so much more manageable that we really get a bargain.

In the master orthogonality relation, set $i = j$, $k = n$, and sum over i, k:

$$\frac{d_\alpha}{{}^oG} \sum_{\substack{s\in G,\\ i,k}} D^{(\alpha)}(s)_{ii} D^{(\beta)}(s^{-1})_{kk} = \sum_{i,k} \delta_{\alpha\beta}\delta_{ik}$$
$$\frac{d_\alpha}{{}^oG} \sum_{s\in G} \chi^{(\alpha)}(s)\chi^{(\beta)}(s^{-1}) = d_\alpha \delta_{\alpha\beta} \tag{9.3}$$

Since problem 9.6 shows us that $\chi(s^{-1}) = \chi(s)^*$ in general (for finite groups) we can finally write

$$\frac{1}{{}^oG} \sum_{s\in G} \chi^{(\alpha)}(s)\chi^{(\beta)}(s)^* = \delta_{\alpha\beta} \tag{9.4}$$

the orthogonality relation for characters of irreducible representations. It is important to remember that our orthogonality relations are based on nonequivalent irreducible representations. From now on we shall reserve the symbols Γ_α, $\mathbf{D}^{(\alpha)}(s)$, $\chi^{(\alpha)}$, and $\chi^{(\alpha)}(s)$ for nonequivalent irreducible representations and Γ, $\mathbf{D}(s)$, χ, and $\chi(s)$ for general (generally reducible) representations.

The character orthogonality expression shows us that the characters of irreducible representations are orthogonal; the reader should verify this for the C_{3v} example presented earlier. Thus, considering Γ_2 and Γ_3

$$\begin{aligned}\langle\chi^{(2)}\,|\,\chi^{(3)}\rangle &\equiv \frac{1}{{}^oG}\sum_{s\in G}\chi^{(2)}(s)^*\chi^{(3)}(s)\\ &= \tfrac{1}{6}\{(1)(2)+(-1)(0)+(-1)(0)+(-1)(0)+(1)(-1)+(1)(-1)\}\\ &= 0\end{aligned} \tag{9.5a}$$

and

$$\begin{aligned}\langle\chi^{(3)}\,|\,\chi^{(3)}\rangle &= \frac{1}{{}^oG}\sum_{s\in G}\chi^{(3)}(s)^*\chi^{(3)}(s)\\ &= \tfrac{1}{6}\{(2)^2+(0)^2+(0)^2+(0)^2+(-1)^2+(-1)^2\}\\ &= 1\end{aligned} \tag{9.5b}$$

We can also reexpress (9.4) in terms of the classes of the group. Let there be $^{\circ}K_i$ ($^{\circ}K_i$ = the order of kth class) elements in the ith class (K_i) which contains the representative element a_i. Since all a_i in that class have the same character, let $\chi(a_i) \equiv \chi_i$ so that (9.4) may be written

$$\frac{1}{^{\circ}G} \sum_i {^{\circ}K_i} \chi_i^{(\alpha)} \chi_i^{(\beta)*} = \delta_{\alpha\beta} \tag{9.4'}$$

where the sum is now over all the nonequivalent classes in G. In C_{3v} the classes are $\{E\}$, $\{\sigma_1, \sigma_2, \sigma_3\}$, and $\{C_3, C_3^2\}$; thus, in this sequence, $^{\circ}K_1 = 1$, $^{\circ}K_2 = 3$, $^{\circ}K_3 = 2$, and 9.5a, b become

$$\langle \chi^{(2)} | \chi^{(3)} \rangle = \tfrac{1}{6}\{1(1)(2) + 3(-1)(0) + 2(1)(-1)\} = 0 \tag{9.5a'}$$

$$\langle \chi^{(3)} | \chi^{(3)} \rangle = \tfrac{1}{6}\{1(2)^2 + 3(0)^2 + 2(-1)^2\} = 1 \tag{9.5b'}$$

Just as we gave a geometrical interpretation to the master orthogonality relations in terms of how many orthogonal vectors we can have in a space of a given dimension, so, too, here we can do the same with our character orthogonality relation. Thus, if C_G is the number of distinct classes in G, Eq. (9.4′) can be interpreted as the scalar product in a C_G-dimensional space where each Γ_α is represented by a vector whose components—one for each class—are

$$\sqrt{\frac{^{\circ}K_i}{^{\circ}G}}\, \chi_i^{(\alpha)}$$

Each irreducible representation contributes one orthogonal vector (also orthonormal in the way we have written them) so that, by the general properties of vector spaces, the number of irreducible representations must be less than or equal to the number of classes of the group. We shall not derive it here but show in Appendix A that the inequality can also be stated in the reverse manner. Taken together, this means that the equality must hold: The number of irreducible representations of a group G is equal to the number of classes in G. We inferred earlier—and shall shortly prove—that the sum of the squares of the dimensions of the irreducible representations is equal to the order of the group

$$\sum_\alpha d_\alpha^2 = {^{\circ}G} \tag{9.6}$$

In addition, it is also true that the dimension of any irreducible representation is a divisor of the order of the group. Taken together these statements allow us to determine, often uniquely, the number and dimensions of the irreducible representations of a group given its order $^{\circ}G$ and the number of classes C_G.

Problem

9.7 Determine the number and dimensions of the irreducible representations of the following groups using the above relations:

a. C_{3v}.
b. *any* abelian group of order $°G$.
c. the group of order 2.
d. D_5, where $°G = 10$ and $C_G = 4$.
e. D_{3h}, where $°G = 12$ and $C_G = 6$.
f. T_d, where $°G = 24$ and $C_G = 5$.

Exercise 9.7 indicates that irreducible representations of dimension one always occur. Actually it should be easy to see that there will *always* be at least one one-dimensional representation for every group. This is so because every group can be represented by the one-dimensional identity representation, $\mathbf{D}(s) = 1$ all $s \in G$, and it is clearly irreducible.

Problem

9.8 Since the identity representation $\Gamma_1 = (\mathbf{D}(s) = 1$, all $s \in G)$ is always one of the irreducible representations of a group, show that the following "sum rule" for characters obtains:

$$\sum_{s \in G} \chi^{(\alpha)}(s) = °G \qquad \Gamma_\alpha = \Gamma_1$$
$$= 0 \qquad \Gamma_\alpha \neq \Gamma_1$$

The derivation that leads to the equality between the number of classes and the number of irreducible representations leads to the result that the class columns of the character table also behave as a set of orthogonal vectors. Thus, if we construct a table of characters by classes, we find *both* row and column orthogonality. As an example, examine the table below for C_{3v}, using the results for the characters and class division noted earlier, where $K_1 = \{E\}$. $K_2 = \{\sigma_1, \sigma_2, \sigma_3\}$, and $K_3 = \{C_3, C_3^2\}$.

C_{3v}	$\chi_1(°K_1 = 1)$	$\chi_2(°K_2 = 3)$	$\chi_3(°K_3 = 2)$
Γ_1	1	1	1
Γ_2	1	-1	1
Γ_3	2	0	-1

In verifying the row and column orthonormality remember the $°K_i$ factors.

CHARACTER TABLES

The example above of a character table is quite general and is the chief way in which we tabulate the information concerning various groups.

Generally the class labels label the columns of the table while the symbols indicating the appropriate irreducible representations label the rows. Since the number of classes equals the number of irreducible representations the table is square. In order to indicate conveniently both the class and the number of elements in the class one usually labels a class column by $°K_i a_i$ where $°K_i$ is the order of the ith class and a_i is *any* one element in the class. Thus, for C_3

C_{3v}	E	3σ	$2C_3$
Γ_1	1	1	1
Γ_2	1	−1	1
Γ_3	2	0	−1

where, for example, σ clearly indicates σ_1, σ_2, or σ_3. The character tables in the appendix can be examined for additional examples.

In addition to this convention regarding class notation there is a standard convention in use for noting irreducible representations in terms of the character properties of that representation. The main letter denotes the dimensionality of the representation while primes and/or additional subscripts (letters or numbers) serve to distinguish otherwise similarly noted representations. The principle C_n axis is always taken to be along the z-direction.

1. One-dimensional representations. The following table gives the basic conventions. The columns are headed by the type of element one considers while the rows are labeled by the value of the character (± 1) of the element.

χ	$C_n^{(z)}$ or $S_n^{(z)}$	σ_v (C_{nv} groups)	$C_2^{(x)}$ (D_n groups)	i	σ_h
1	A	subscript 1	subscript 1	subscript g	superscript single prime
−1	B	subscript 2	subscript 2	subscript u	superscript double prime

For D_2 and D_{2h} where there are three twofold axes, B_1, B_2, and B_3 are used when $a+1$ character is associated with $C_2^{(z)}$, $C_2^{(y)}$, and $C_2^{(x)}$, respectively; in each case $a-1$ character is associated with the *other two* elements. A_1 is used only for the representation where all characters are $+1$. The subscripts g (gerade) and u (ungerade) are the German equivalent of even and odd, respectively.

2. Two-dimensional representations. These are denoted by E. When necessary g, u, and/or single or double prime notation is added according

to the character of the operations i and/or σ_h. If these are insufficient or unavailable, we shall find that the character associated with C_n of the group will be $2\cos(2\pi p/n)$, and the subscript $p = 1, 2, \ldots$ is then employed. Although *not* two-dimensional, for reasons that will be discussed later, representations that are complex conjugates of each other are usually written as a pair and are also denoted by E; in such cases we shall find $\chi(C_n) = e^{i2\pi p/n}$, where p is used as a subscript to identify the various pairs of one-dimensional representations as well as primes and/or g or u, where applicable.

3. Higher dimensional representations. Three-, four-, and five-dimensional representations are denoted by T, G, and H, respectively. They are of less general occurrence, and since their subscription is less direct than that for lower dimensional representations, we do not attempt to define it. G and H representations occur only in I and I_h (the icosahedral groups) as do also T representations. The other T representations occur in the cubic groups T, T_d, T_h, O, and O_h.

The reader can familiarize himself with the various conventions by studying the group tables in the appendix and by employing this notation in our subsequent discussions.

When describing the symmetry properties of a system it is usually a good idea to define the coordinate system. Among chemists there is a convention for choosing coordinate systems for molecules that bears mentioning; basically it is a convention that attempts to make optimum use of the symmetry present. The principle axis (of highest symmetry) is, as noted before, always chosen as the z-axis in a right-handed coordinate system. If we have a choice of the principle axis (such as, for example, D_2) we choose the z-axis to pass through the largest number of atoms. The x-axis choice occurs for planar molecules; if the z-axis is in the molecular plane the x-axis is taken perpendicular to the molecular plane, while if the z-axis is normal to the plane of the molecule, the x-axis is chosen to pass through the most atoms in the plane. The reader should remember that this is simply a *convention;* the choice of a particular coordinate system can never affect the physical results of a problem. In general it is always a good idea to carefully specify the coordinate system employed.

THE DECOMPOSITION OF REDUCIBLE REPRESENTATIONS

Suppose we are given a representation Γ which may be reducible. If it is, then by a suitable transformation it can be reduced to the point where it can be written as a direct sum of irreducible representations of the group in question.

$$\Gamma = \Gamma_1 \oplus \Gamma_2 \oplus \cdots = \sum_\alpha c_\alpha \Gamma_\alpha \tag{9.7}$$

where c_α tells us how many times Γ_α occurs in (the transformed) Γ. Now although the above equation is a symbolic sum, an equation for the character of a particular element s can be written which is a "true" algebraic relation:

$$\chi(s) = \sum_\alpha c_\alpha \chi^{(\alpha)}(s) \tag{9.8}$$

That this is true is readily apparent from what is meant by the sum of representations. Note, too, that the character equation is independent of any and all similarity transformations that can be applied to the original untransformed and unreduced representation. Given now the characters of the various irreducible representations of G, we can determine c_β by multiplying $\chi(s)$ by $\chi^{(\beta)}(s)^*$ and summing over the group elements s. For, by the character orthogonality relation,

$$\begin{aligned} \sum_{s\in g} \chi^{(\beta)}(s)^* \chi(s) &= \sum_{s\in g} c_\alpha \chi^{(\beta)}(s)^* \chi^{(\alpha)}(s) \\ &= c_\beta {}^oG \end{aligned} \tag{9.9}$$

so that

$$\begin{aligned} c_\beta &= \frac{1}{{}^oG} \sum_{s\in g} \chi^{(\beta)}(s)^* \chi(s) \\ &= \frac{1}{{}^oG} \sum_i {}^oK_i \chi_i^{(\beta)*} \chi_i \end{aligned} \tag{9.10}$$

the last two equations simply being the sum over elements or the weighted (by ${}^oK_i$) sum over classes. By a very simple process, then, we can deduce the composition of any representation of a group in terms of its irreducible representations.

The process involved is quite analogous to Fourier analysis or the determination in quantum mechanics of the composition of an arbitrary wave function $|\phi\rangle$ in terms of some orthonormal basis.

$$\begin{aligned} |\phi\rangle &= \sum_n c_n |n\rangle \\ c_m &= \langle m \mid \phi\rangle \end{aligned} \tag{9.11}$$

That is to say, we may interpret χ (the *set* $\{\chi(s)\}$) as a vector in the C_G-dimensional class space of G and find its composition in terms of the C_G orthonormal $\chi_i^{(\alpha)}$ vectors by developing the appropriate (complex) scalar product.

There is a convenient way to carry out such an analysis using the form of the character table of the pertinent group. For example, suppose there is a representation of C_{3v} with the characters $\chi(E) = 3$, $\chi(C_3) = 0$, and

$\chi(\sigma)=1$. Add this representation to the character table as shown below and note that the necessary scalar product is simply given by multiplying the appropriate Γ and Γ_α rows using the ${}^oK_i$ weighting factors.

C_{3v}	E	$2C_3$	$3\sigma_v$
A_1	1	1	1
A_2	1	1	−1
E	2	−1	0
Γ	3	0	1

$$\begin{aligned} c_{A_1} &= \frac{1}{G}[1(1)(3)+2(1)(0)+3(1)(1)]=1 \\ c_{A_2} &= \frac{1}{G}[1(1)(3)+2(1)(0)+3(-1)(1)]=0 \\ c_E &= \frac{1}{G}[1(2)(3)+2(-1)(0)+3(0)(1)]=1 \\ \Gamma &= A_1+E \end{aligned} \tag{9.12}$$

Thus, Γ is composed of the A_1 and E irreducible representations, each occurring once.

It turns out that representations of a group G formed by taking the direct product of irreducible representations of the direct factors of G are irreducible. That is, for $G=H\times K$ and given the irreducible representations of H and K, $\Gamma_\alpha^{(H)}$, $\Gamma_\beta^{(K)}$, then all the nonequivalent irreducible representations of G are given by

$$\Gamma_{(\alpha\beta)}^{(G=H\times K)}=\Gamma_\alpha^{(H)}\otimes\Gamma_\beta^{(K)} \quad (\alpha,\beta=1,2,\ldots) \tag{9.13}$$

This follows from the fact that

$$\chi^{(\alpha\beta;G)}(s)=\chi^{(\alpha;H)}(s)\chi^{(\beta;K)}(s) \tag{9.14}$$

and that a representation is irreducible if

$$\langle\chi\,|\,\chi\rangle\equiv\frac{1}{{}^oG}\sum_{s\in G}\chi(s)^*\chi(s)=1 \tag{9.15}$$

the character normalization condition (see exericse 9.9 below). Such is generally not true for direct products *within* a group. Thus,

$$\begin{aligned} \Gamma &= \Gamma_\alpha\times\Gamma_\beta \\ \chi(s) &= \chi^{(\alpha)}(s)\chi^{(\beta)}(s) \end{aligned} \tag{9.16}$$

As an example, for C_{3v} the character of $E\times E$ is $\chi(E)=4$, $\chi(C_3)=1$,

$\chi(\sigma)=0$, and by our decomposition procedure we find

$$E\times E = A_1 + A_2 + E \tag{9.17}$$

On the other hand,

$$\begin{aligned} A_1\times A_2 &= A_2 \\ A_2\times A_2 &= A_1 \\ E\times A_2 &= E \end{aligned} \tag{9.18}$$

as may be verified by the reader. We can, of course, extend this to direct products of as many representations, irreducible or reducible, as we like.

Problems

9.9 Show that a representation is irreducible if

$$\langle\chi\,|\,\chi\rangle = \frac{1}{{}^{o}G}\sum_{s\in G}\chi(s)^*\chi(s) = 1$$

9.10 Find the composition of a representation of C_{3v} where

$$\chi(E)=9, \qquad \chi(C_3)=0, \qquad \chi(\sigma)=-1$$

9.11 Show that the direct product within a group of any two one-dimensional representations is irreducible.

9.12 For the group C_{4v} a representation has characters $\chi(E)=5$, $\chi(C_4)=-1$, $\chi(C_2)=1$, $\chi(\sigma_v)=1$, and $\chi(\sigma_\alpha)=-3$. Resolve this representation into its irreducible components.

9.13 For C_{4v}, resolve

a. $E\otimes E$

b. $A_2\otimes A_2\otimes E$

c. $B_1\otimes B_2$

9.14 Consider C_3 to consist of three one-dimensional irreducible representations. Find all the direct products within the group.

The reader will have noticed some generalizations in the above problems. First, and perhaps most obvious, is the check on the decomposition of a representation that the sum of the dimensions of the irreducible representations must equal the dimension of the original representation. Equally obvious is the fact that $\Gamma_1\times\Gamma_\alpha=\Gamma_\alpha$, where we denote Γ_1 as the totally symmetric representation.

THE REGULAR REPRESENTATION

The regular representation is an interesting one because of the special properties of the permutation group isomorphic with it. Recall that Cayley's theorem tells us that every group of order n is isomorphic to some subgroup of the symmetric group S_n. In particular we found that the permutation subgroup isomorphic with a particular group G consisted of so-called regular permutations that (aside from the unit element) leave no object in an ordered array unchanged. The matrix representation of the identity element e in the regular representation Γ_R of G is simply

$$\mathbf{D}^R(e)=\begin{pmatrix} 1 & 0 & 0 & \cdots \\ 0 & 1 & 0 & \cdots \\ 0 & 0 & 1 & \cdots \\ \cdot & & & \\ \cdot & & & \\ \cdot & & & \end{pmatrix} \tag{9.19}$$

as usual, and the matrix representations have *all* the rows of $\mathbf{D}^R(e)$ interchanged according to the permutation involved (see Chapter 7). It follows that

$$\begin{aligned} \chi^R(e) &= d_R = {}^{\circ}G \\ \chi^R(s) &= 0, \qquad s \neq e, \qquad s \in G \end{aligned} \tag{9.20}$$

for the regular representation of a group G of order ${}^{\circ}G$.

Let Γ_α, Γ_β be the irreducible representations of G. Then we can analyze the structure of Γ_R in the usual way.

$$\begin{aligned} \Gamma_R &= \sum_\alpha c_\alpha \Gamma_\alpha \\ c_\alpha &= \frac{1}{{}^{\circ}G} \sum_s \chi^{(\alpha)}(s)^* \chi^R(s) \\ &= \frac{1}{{}^{\circ}G} \chi^{(\alpha)}(e)^* \chi^R(e) \\ &= \frac{1}{{}^{\circ}G} d_\alpha^{\circ} G = d_\alpha \end{aligned} \tag{9.21}$$

and we see why the regular representation is sometimes referred to as the master representation: The regular representation contains each irreducible representation a number of times equal to the dimension of the irreducible representation. This property can be used to prove another equality alluded to previously. Thus,

$$\langle \chi^R \mid \chi^R \rangle = \frac{1}{{}^{\circ}G} \sum_s \chi^R(s)^* \chi^R(s) = \frac{1}{{}^{\circ}G} ({}^{\circ}G)^2 = {}^{\circ}G \tag{9.22}$$

since only $\chi^R(e)$ is nonzero and is simply ${}^{\circ}G$. But also,

$$\begin{aligned} \langle \chi^R \mid \chi^R \rangle &= \frac{1}{{}^{\circ}G} \sum_{\alpha,\beta,s} c_\alpha^* c_\beta \chi^{(\alpha)}(s)^* \chi^{(\beta)}(s) \\ &= \sum_\alpha c_\alpha^2 = \sum_\alpha d_\alpha^2 \end{aligned} \tag{9.23}$$

where the last equality is just what was proved above. Thus, equating (9.22) and (9.23),

$$°G = \sum_{\alpha} d_{\alpha}^2 \tag{9.24}$$

and we have proved that the sum of the squares of the irreducible representations of a group is equal to the order of the group.

ABELIAN AND CYCLIC GROUPS; COMPLEX CONJUGATE CHARACTERS

As has been stated previously, one-dimensional representations and their characters are one and the same since $\mathbf{D}^{(\alpha)}(s) = \chi^{(\alpha)}(s)$ when Γ_α is one-dimensional. In particular, when a group is limited to only one-dimensional representations the character table is particularly easy to construct. Such is the case when we deal with abelian groups; for abelian groups all elements commute so that since each element is in a class by itself there are exactly $°G$ one-dimensional representations. Now characters for one-dimensional representations cannot vanish and must be of modulus 1. Thus, the construction of the character table by forming $°G$ orthogonal vectors in the $°G$-dimensional class space is relatively straightforward.

Problem

9.15 Show that $\chi(s) = (e^{2\pi i/n})^m$ ($m = 1, 2, \ldots, n$ or $m = 0, 1, 2, \ldots, n-1$) are the only allowable characters in one-dimensional representations where s is an element of order n.

9.16 Show that groups $G = \{e, a, b, \ldots\}$ where $s^2 = e$, *all* $s \in G$, must be abelian. (Use the fact that if $ab = c$, $e = c^2 = (ab)^2$.)

9.17 Use the results of the preceding two problems to construct the character table of the point group C_{2v}.

Because it arises in many situations of physical interest, the cyclic (and, thus, abelian) group and its character table deserve special mention. Cyclic symmetry is usually easy to recognize. The benzene molecule has a C_6-axis and, although it has a higher symmetry (D_{6h}), we can classify many of its interesting properties by considering its cyclic symmetry alone. A one-dimensional lattice of N equidistant points has a C_2-axis bisecting the line of lattice points; if we presume cyclic boundary conditions (physically permissable on the limit of large N), it is the same as if one end of the lattice is connected to the other and we can immediately treat the system as though it were a circle of N points; that is, it becomes equivalent to a system with C_N symmetry. As we shall see later, this allows us to write down, for example, the correct single particle wave functions for such systems by utilizing the symmetry properties alone.

It is easy to see by extending the argument presented in problem 9.15 that for the cyclic group (isomorphic to C_n) generated by the element s of order $n = {}^{\circ}G$ (since $s^n = e$)

$$\chi(s) = (e^{2\pi i/n})^m$$
$$\equiv \varepsilon^m, m = 1, 2, \ldots, n-1, n \tag{9.25}$$

are the possible characters of s. Since the group is cyclic $\chi(s^2) = [\chi(s)]^2$ and, in general, $\chi(s^k) = [\chi(s)]^k = \varepsilon^{mk}$ $(k = 1, 2, \ldots, n)$. We note that the n choices of m and the n choices of k give us exactly n^2 characters to fill the character table. For the moment let the value of m denote the representation in question. Then the character table takes the form shown in Table 1. The last row (Γ_n) follows from the fact that ε^n and all its powers are unity, Γ_n being the identity representation present in *all* groups. We note that the representations occur in pairs in the sense that Γ_k and Γ_{n-k} are complex conjugates of one another.

$$\chi^{(k)}(s) = \varepsilon^k = \varepsilon^{k-n} = (\varepsilon^{n-k})^* = \chi^{(n-k)}(s)^* \tag{9.26}$$

Table 1

C_n	e	s	s^2	$\cdots$	s^{n-1}
Γ_1	1	ε	ε^2	$\cdots$	$\varepsilon^{(n-1)}$
Γ_2	1	ε^2	ε^4	$\cdots$	$\varepsilon^{2(n-1)}$
$\cdot$					
$\cdot$					
$\cdot$					
Γ_{n-1}	1	ε^{n-1}	$\varepsilon^{2(n-1)}$	$\cdots$	$\varepsilon^{(n-1)(n-1)}$
Γ_n	1	1	1	$\cdots$	1

Γ_n is conventionally denoted as A while the conjugate representations Γ_k and Γ_{n-k} are normally bracketed together and denoted as E_k.

Problem

9.18 Construct the character tables for the cyclic groups of orders 3, 4, and 5.

The one-dimensional representations paired together as E, denoting a two-dimensional irreducible representation, are legitimate one-dimensional irreducible representations for the point groups in which they are noted. As we shall discuss later, in quantum mechanical applications it turns out that the dimensionality of an irreducible representation present in a physical system tells us the energy degeneracy associated with the wave functions that carry (provide a basis for) the irreducible representation. For this degeneracy–dimensionality relation to be observed one must choose the *full* symmetry group of the Hamiltonian. Although we usually think of the proper symmetry groups in terms of spatial operations, often the correct group is really the direct-product group consisting of the spatial group and the 2-element group made up of e and the element representing the operation of time inversion. Such will always be the case when the Hamiltonian is real. Rather than work with the full group which includes the time inversion operator, it is simpler to work with the spatial subgroup and a set of rules that tell us when "extra" degeneracy will arise if we employ the full group. The representations whose characters are complex conjugates of each other will be degenerate (in the quantum mechanical energy sense) when a physical system is invariant under time inversion. This is, then, the reason why such representations are usually paired together and denoted as E types. The labeling of these special one-dimensional representations does not affect in any way the manipulations we perform with them later on; it simply serves to tell us that we can anticipate a result (extra energy degeneracy) we might not otherwise expect.

SOLUTIONS TO PROBLEMS

9.1
$$\begin{aligned}\mathrm{Tr}\,\mathbf{ABC} &= \sum_{ijk} A_{ij}B_{jk}C_{ki}\\ &= \sum_{ijk} C_{ki}A_{ij}B_{jk} = \mathrm{Tr}\,\mathbf{CAB}\\ &= \sum_{ijk} B_{jk}C_{ki}A_{ij} = \mathrm{Tr}\,\mathbf{BCA}\end{aligned}$$

9.2 If $\mathbf{A} = \mathbf{U}^{-1}\mathbf{BU}$, then by 9.1

$$\mathrm{Tr}\,\mathbf{A} = \mathrm{Tr}\,(\mathbf{U}^{-1}\mathbf{BU}) = \mathrm{Tr}\,(\mathbf{UU}^{-1}\mathbf{B}) = \mathrm{Tr}\,\mathbf{B}$$

9.3 a. $(\mathbf{A}\otimes\mathbf{B})_{ij,kl} = A_{ik}B_{jl}$

$$\begin{aligned}\mathrm{Tr}\,(\mathbf{A}\otimes\mathbf{B}) &= \sum_{ij} (\mathbf{A}\otimes\mathbf{B})_{ij,ij}\\ &= \sum_{ij} A_{ii}B_{jj} = (\mathrm{Tr}\,\mathbf{A})(\mathrm{Tr}\,\mathbf{B})\end{aligned}$$

b. $\mathbf{A}\oplus\mathbf{B} = \begin{pmatrix}\mathbf{A} & 0\\ 0 & \mathbf{B}\end{pmatrix}$

so clearly
$$\begin{aligned}\mathrm{Tr}\,(\mathbf{A}\oplus\mathbf{B}) &= \sum_i A_{ii} + \sum_j B_{jj}\\ &= \mathrm{Tr}\,\mathbf{A} + \mathrm{Tr}\,\mathbf{B}\end{aligned}$$

9.4 Two elements a and b, belonging to the same class, are related by

$a = sbs^{-1}$ for some $s \in G$, so that

$$\mathbf{D}(a) = \mathbf{D}(s)\mathbf{D}(b)\mathbf{D}(s^{-1})$$

Thus

$$\begin{aligned}\operatorname{Tr} \mathbf{D}(a) &= \operatorname{Tr} \{\mathbf{D}(s)\mathbf{D}(b)\mathbf{D}(s^{-1})\} \\ &= \operatorname{Tr} \{\mathbf{D}(s^{-1})\mathbf{D}(s)\mathbf{D}(b)\} \\ &= \operatorname{Tr} \mathbf{D}(b)\end{aligned}$$

Confirm this result by comparing the characters in a particular irreducible representation of the elements in the same classes in C_{3v} as given in the text.

9.5 For a representation of dimension d, $\mathbf{D}(e) = \mathbf{1}$, where $\mathbf{1}$ is the unit matrix of dimension d. Thus,

$$\operatorname{Tr} \mathbf{D}(e) = \operatorname{Tr} \mathbf{1} = \sum_{i}^{d} 1 = d$$

9.6 Theorem 1 in Chapter 8 stated that for finite groups any representation is equivalent to a unitary representation. That is, if Γ is not unitary, there exists a transformation such that the set Γ' is unitary.

$$\mathbf{D}'(s) = \mathbf{U}\mathbf{D}(s)\mathbf{U}^{-1}$$

$$\mathbf{D}'(s)^{-1} = \mathbf{D}'(s)^{\dagger}$$

Thus,

$$\chi(s) = \operatorname{Tr} \mathbf{D}(s) = \operatorname{Tr} \mathbf{U}^{-1}\mathbf{D}'(s)\mathbf{U} = \operatorname{Tr} \mathbf{D}'(s)$$

so

$$\chi(s^{-1}) = \operatorname{Tr} \mathbf{D}(s^{-1}) = \operatorname{Tr} \mathbf{D}'(s^{-1})$$

$$\mathbf{D}'(s^{-1}) = \mathbf{D}'(s)^{-1} = \mathbf{D}'(s)^{\dagger}$$

$$\begin{aligned}\chi(s^{-1}) &= \operatorname{Tr} \mathbf{D}'(s)^{\dagger} = \sum_{i} D'(s)^{*}_{ii} \\ &= (\operatorname{Tr} \mathbf{D}'(s))^{*} \\ &= (\operatorname{Tr} \mathbf{D}(s))^{*} = \chi(s)^{*}\end{aligned}$$

Note that the result $\chi(s^{-1}) = \chi(s)^{*}$ has been shown to be valid (for finite groups) whether Γ is unitary or not.

9.7 a. By now we know that for C_{3v}, $^{o}G = 6$ and $C_G = 3$. Thus, there are three irreducible representations and

$$d_1^2 + d_2^2 + d_3^2 = 6$$

whose *only* solution is $d_1 = 1$, $d_2 = 1$, $d_3 = 2$.

b. In an abelian group each element is in a class by itself; thus $c_G = {^{o}G}$ and

$$\sum_{\alpha=1}^{^{o}G} d_\alpha^2 = {^{o}G}$$

requires *every* $d_\alpha = 1$. Thus, for all abelian groups there are only one-dimensional representations.

c. The group of order 2 is obviously abelian. There are, thus, two classes and two one-dimensional representations.

d. $C_G = 4$ and $^\circ G = 10$.

$$d_1^2 + d_2^2 + d_3^2 + d_4^2 = 10$$

The only solution is $d_1 = d_2 = 1$ and $d_3 = d_4 = 2$.

e. $C_G = 6$ and $^\circ G = 12$.

$$\sum_\alpha d_\alpha^2 = 12$$

and the solution is $d_1 = d_2 = d_3 = d_4 = 1$ and $d_5 = d_6 = 2$.

f. $C_G = 5$, $^\circ G = 24$.

$$\sum_\alpha d_\alpha^2 = 24$$

and the solution is $d_1 = d_2 = 1$, $d_3 = 2$, and $d_4 = d_5 = 3$.

9.8 This result follows readily from the character orthogonality relation

$$\sum_{s \in G} \chi^{(\alpha)}(s)^* \chi^{(\beta)}(s) = {}^\circ G \delta_{\alpha\beta}$$

Simply set $\alpha = 1$, the identity representation; since $\chi^{(\alpha)}(s) = 1$, all s, we get

$$\sum_{s \in G} \chi^{(\beta)}(s) = {}^\circ G \delta_{1\beta}$$

9.9 If $\Gamma = \sum_\alpha c_\alpha \Gamma_\alpha$ is irreducible then one and only one c_α must equal unity. If χ is the character of Γ and $\Gamma = \Gamma_\alpha$, then

$$c_\beta = \frac{1}{{}^\circ G} \sum_{s \in G} \chi^{(\beta)}(s)^* \chi^{(\alpha)}(s) = \delta_{\alpha\beta}$$

Or, if generally

$$\Gamma = \sum_\alpha c_\alpha \Gamma_\alpha, \ \chi(s) = \sum_\alpha c_\alpha \chi^{(\alpha)}(s)$$

$$\langle \chi \mid \chi \rangle = \frac{1}{{}^\circ G} \sum_{\alpha,\beta,s} c_\alpha^* c_\beta \chi^{(\alpha)}(s)^* \chi^{(\beta)}(s)$$

$$= \sum_\alpha c_\alpha^* c_\alpha$$

which can equal unity only if one $c_\beta = 1$ and all the others equal zero.

9.10

C_{3v}	E	$2C_3$	3σ	
A_1	1	1	1	$c_{A_1}=1$
A_2	1	1	−1	$c_{A_2}=2$
E	2	−1	0	$c_E=3$
Γ	9	0	−1	

$$\Gamma = A_1 + 2A_2 + 3E$$

9.11 Since $\chi(s) = \chi^{(\alpha)}(s)\chi^{(\beta)}(s)$ for any direct product, if Γ_α and Γ_β are one-dimensional so, too, will $\Gamma_\alpha \otimes \Gamma_\beta$. Since *any* one-dimensional representation is irreducible, the direct product of any two one-dimensional representations will also be irreducible.

9.12

C_{4v}	E	$2C_4$	C_2	$2\sigma_v$	$2\sigma_d$	
A_1	1	1	1	1	1	$c_{A_1}=0$
A_2	1	1	1	−1	−1	$c_{A_2}=1$
B_1	1	−1	1	1	−1	$c_{B_1}=2$
B_2	1	−1	1	−1	1	$c_{B_2}=0$
E	2	0	−2	0	0	$c_E=1$
Γ	5	−1	1	1	−3	

$$\Gamma = A_2 + 2B_1 + E$$

9.13 Referring to the above table for C_{4v} we find the following characters:

	E	$2C_4$	C_2	$2\sigma_v$	$2\sigma_d$
$E \otimes E$	4	0	4	0	0
$A_2 \otimes A_2 \otimes E$	2	0	−2	0	0
$B_1 \otimes B_2$	1	1	1	−1	−1

and

$$E \otimes E = A_1 + A_2 + B_1 + B_2$$

$$A_2 \otimes A_2 \otimes E = E$$

$$B_1 \otimes B_2 = A_2$$

9.14

	C_3	E	C_3	C_3^2
A	Γ_1	1	1	1
E	Γ_2	1	ε	ε^*
	Γ_3	1	ε^*	ε

$$°G = 3$$

$$\varepsilon = e^{2\pi i/3}$$

$$\varepsilon^2 = e^{4\pi i/3} = e^{-2\pi i/3} = \varepsilon^*$$

$$(\varepsilon^*)^2 = \varepsilon$$

$$\Gamma_1 \otimes \Gamma_1 = \Gamma_1 \qquad \Gamma_2 \otimes \Gamma_2 = \Gamma_3$$
$$\Gamma_1 \otimes \Gamma_2 = \Gamma_2 \qquad \Gamma_2 \otimes \Gamma_3 = \Gamma_1$$
$$\Gamma_1 \otimes \Gamma_3 = \Gamma_3 \qquad \Gamma_3 \otimes \Gamma_3 = \Gamma_2$$

Note that if we write $E = \Gamma_2 + \Gamma_3$, $\varepsilon + \varepsilon^* = 2\cos(2\pi/3) = -1$

C_3	E	C_3	C_3^2
A	1	1	1
E	2	−1	−1

the row character for E is not properly normalized ($\langle E \mid E \rangle \neq 1$) and the usual orthogonality rules will not always work. Thus, while, as in the above example of C_3, we use E symbolically, it is important to write each irreducible representation out explicitly to remind us of their true one-dimensional nature.

9.15 Since s is of order n, $[\mathbf{D}(s)]^n = \mathbf{D}(e) = \mathbf{1}$ where $\mathbf{1}$ is the unit matrix of dimension equal to the dimension of the representation. For one-dimensional representations the equation becomes one for the characters

$$[\chi(s)]^n = \chi(e) = 1$$

so that $\chi(s)$ must be one of the n roots of unity

$$\chi(s) = (e^{2\pi i/n})^m, \qquad m = 1, 2, \ldots, n$$

9.16 Let $ab = c$ where $a^2 = b^2 = c^2 = e$. Then $(ab)^2 = abab = c^2 = e$ and

$$abab = e$$
$$aba = b$$
$$ab = ba$$

by multiplying on the right by b and then a and making use of the fact that $a^2 = b^2 = e$. Thus $ab = ba$ must be true for all pairs of elements in the group and the group is abelian.

9.17 $C_{2v} = \{E, C_2, \sigma_v, \sigma_{v'}\}$ where each element is of order 2. The group is abelian and thus must have four one-dimensional irreducible representations where $\chi(s) = \pm 1$ for $n = 2$ (see problem 9.15). We automatically know the part of the table above and to the left of the dashed lines indicated below; the remainder is

C_{2v}	E	C_2	σ_v	$\sigma_{v'}$
A_1	1	1	1	1
A_2	1	1	−1	−1
B_1	1	−1	1	−1
B_2	1	−1	−1	1

$$\sigma_v = \sigma^{(xz)}$$
$$\sigma_{v'} = \sigma^{(yz)}$$

readily filled in by remembering that $\sum_s \chi^{(\alpha)}(s) = 0$ for $\Gamma_\alpha \neq \Gamma_{\text{identity}}$ (completely symmetric, identity representation) and by forming four orthogonal class vectors.

9.18 Each group is isomorphic to C_n with $n = 3, 4, 5$. For each case ε is defined and the table presented is in the general form shown in the text. To the right is given the form of the table as it is usually found in the tables, where E representations are listed together and simplifying features of the generally complex nature of ε are used.

C_3	E	C_3	C_3^2
Γ_1	1	ε	ε^2
Γ_2	1	ε^2	ε
Γ_3	1	1	1

C_3	E	C_3	C_3^2
A	1	1	1
E	1	ε	ε^*
	1	ε^*	ε

C_4	E	C_4	$C_4^2 = C_2$	C_4^3
Γ_1	1	i	-1	$-i$
Γ_2	1	-1	1	-1
Γ_3	1	$-i$	-1	i
Γ_4	1	1	1	1

$\varepsilon = e^{2\pi i/4} = i$

C_4	E	C_4	C_2	C_4^3
A	1	1	1	1
B	1	-1	1	-1
E	1	i	-1	$-i$
	1	$-i$	1	i

C_5	E	C_5	C_5^2	C_5^3	C_5^4
Γ_1	1	ε	ε^2	ε^3	ε^4
Γ_2	1	ε^2	ε^4	ε^6	ε^8
Γ_3	1	ε^3	ε^6	ε^9	ε^{12}
Γ_4	1	ε^4	ε^8	ε^{12}	ε^{15}
Γ_5	1	1	1	1	1

$\varepsilon = e^{2\pi i/5}$

	E	C_5	C_5^2	C_5^3	C_5^4
A	1	1	1	1	1
E_1	1	ε	ε^2	ε^{*2}	ε^*
	1	ε^*	ε^{*2}	ε^2	ε
E_2	1	ε^2	ε^*	ε	ε^{*2}
	1	ε^{*2}	ε	ε^*	ε^2

10
The Transformation of Functions and Coordinates

THE TRANSFORMATION OF COORDINATES

In discussing representations and their construction we have used the idea of specifying a coordinate system, finding a set of vectors, and then defining the operations associated with the group elements on the basis vectors of the carrier space. We have alluded to the fact that functions in a function space can serve the same purpose as "ordinary" basis vectors although no specific examples have been employed. In quantum mechanics we are interested in the transformation properties of wave functions and how group theory can help us in their manipulation. In this chapter and the next we shall discuss functions as carriers (basis vectors) of representations and how one goes about selecting or forming such a set (a symmetry adapted basis set).

In discussing how functions are rotated, inverted, reflected, or generally transformed by an operator of a group we need to know how coordinates themselves transform. Since most often we are involved with spatial symmetry operations, we concentrate on point group operations and their effect on Cartesian coordinates of a point in ordinary three-dimensional space. Cartesian coordinates are especially convenient to work with since their use implies an orthogonal basis (unit metric tensor).

Given three (independent) unit vectors defining a "real-space" coordinate system the action of any point group operation $\hat{a}$ on one of the three basis vectors will be given by

$$\hat{a}\vec{\varepsilon}_i = \sum_j \vec{\varepsilon}_j A_{ji} \quad i = 1, 2, 3 \tag{10.1}$$

or in terms of the row vector of basis vectors

$$\boldsymbol{\varepsilon} = (\vec{\varepsilon}_1, \vec{\varepsilon}_2, \vec{\varepsilon}_3) \tag{10.2}$$

$$\hat{a}\boldsymbol{\varepsilon} = \boldsymbol{\varepsilon}\mathbf{A}$$

where $\mathbf{A}$ is the matrix representative (the vector representative) of $\hat{a}$ in the basis defined by $\boldsymbol{\varepsilon}$. We are repeating the exact procedure developed in Chapter 7, where the basic definition of a representation was introduced.

For each operator $\hat{a} \in G$ there will be a matrix **A** and the matrix set will form a representation of G

$$\begin{aligned} \hat{a}\hat{b} &= \hat{c} \\ \hat{c}\boldsymbol{\varepsilon} &= \boldsymbol{\varepsilon}\mathbf{C} \\ \hat{a}\hat{b}\boldsymbol{\varepsilon} &= \hat{a}\boldsymbol{\varepsilon}\mathbf{B} = \boldsymbol{\varepsilon}\mathbf{A}\mathbf{B} \\ \mathbf{C} &= \mathbf{A}\mathbf{B} \end{aligned} \tag{10.3}$$

Since we are here restricting ourselves to a three-dimensional space we will generate a three-dimensional representation.

Given $\boldsymbol{\varepsilon}$ any vector in our ordinary space can be written as

$$\begin{aligned} \vec{v} &= \boldsymbol{\varepsilon}\mathbf{v} \\ &= \sum_i \vec{\varepsilon}_i \mathbf{v}_i \end{aligned} \tag{10.4}$$

where $\mathbf{v}$ is the column vector of the coefficient of the $\vec{\varepsilon}_i$ in $\vec{v}$. The action of $\hat{a}$ on $\vec{v}$ will then be given by

$$\begin{aligned} \hat{a}\vec{v} &= \hat{a}\boldsymbol{\varepsilon}\mathbf{v} \\ &= \boldsymbol{\varepsilon}\mathbf{A}\mathbf{v} \\ &= \boldsymbol{\varepsilon}\mathbf{v}' \\ \mathbf{v}' &= \mathbf{A}\mathbf{v} \end{aligned} \tag{10.5}$$

Although the construction of representations per se is usually done more easily by determining the operation of the various $\hat{a} \in G$ on the basis vectors, it can also easily be done by the equivalent process of considering the transformation $\mathbf{v} \rightarrow \mathbf{v}'$ and finding the desired matrix that effects it.

Thus, consider a vector $\vec{r}$ in a Cartesian coordinate system with components x, y, z represented by

$$\mathbf{r} = \begin{pmatrix} x \\ y \\ z \end{pmatrix}$$

and a set of operators $\hat{O}_a \in G$ and their representative matrices $\mathbf{D}(\hat{O}_a)$. What happens to the components of the vector $\vec{r}$ upon inversion through the origin by the operation $\hat{O}_i$? Clearly

$$\begin{pmatrix} x \\ y \\ z \end{pmatrix} \rightarrow \begin{pmatrix} -x \\ -y \\ -z \end{pmatrix} \tag{10.6}$$

We see, therefore, that there exists a matrix representation of $\hat{O}_i$, $\mathbf{D}(i)$, where in this example

$$\begin{pmatrix} -x \\ -y \\ -z \end{pmatrix} = \mathbf{D}(i)\begin{pmatrix} x \\ y \\ z \end{pmatrix} = \begin{pmatrix} -1 & 0 & 0 \\ 0 & -1 & 0 \\ 0 & 0 & -1 \end{pmatrix}\begin{pmatrix} x \\ y \\ z \end{pmatrix} \tag{10.7}$$

What is the effect of a rotation by θ degrees about the z-axis on a vector in the x–y-plane? We shall define the rotation to be clockwise looking down the positive z-axis as indicated in Figure 1. To find the appropriate

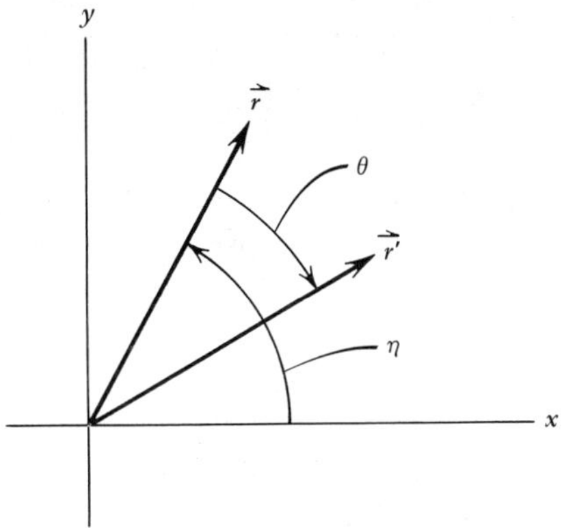

Figure 1

relation it proves convenient to define η as the angle $\vec{r}$ makes with the x-axis. Then, $x = r\cos\eta$, $y = r\sin\eta$ and we find for x' and y'

$$\begin{aligned} x' &= r\cos(\eta-\theta) = r\cos\eta\cos\theta + r\sin\eta\sin\theta \\ &= x\cos\theta + y\sin\theta \\ y' &= r\sin(\eta-\theta) = r\sin\eta\cos\theta - r\cos\eta\sin\theta \\ &= y\cos\theta - x\sin\theta \end{aligned} \tag{10.8}$$

where r is the magnitude of $\vec{r}$ and is invariant under *any* point group operation centered at the point of the origin. Writing the result in matrix notation

$$\begin{pmatrix} x' \\ y' \end{pmatrix} = \begin{pmatrix} \cos\theta & \sin\theta \\ -\sin\theta & \cos\theta \end{pmatrix}\begin{pmatrix} x \\ y \end{pmatrix} \tag{10.9}$$

we easily extend this particular operation to a general vector (x, y, z) by recognizing that under any rotation about the z-axis, the z-coordinate does not change.

$$D(C_z(\theta)) = \begin{pmatrix} \cos\theta & \sin\theta & 0 \\ -\sin\theta & \cos\theta & 0 \\ 0 & 0 & 1 \end{pmatrix} \tag{10.10}$$

The transformation properties of other operators can be similarly derived. With each operator $\theta_t \in G$ can be associated a matrix **T** such that $\mathbf{r}' = \mathbf{Tr}$ and such that the set $\{\mathbf{T}\}$ forms a three-dimensional (usually reducible) representation of G.

Problems

10.1 a. Use the methods above to show that the transformation matrix for reflection through a vertical plane (a reflection plane containing the z-axis) that makes an angle of θ degrees with the x-axis is

$$\mathbf{D}(\sigma_v(\theta)) = \begin{pmatrix} \cos 2\theta & \sin 2\theta & 0 \\ \sin 2\theta & -\cos 2\theta & 0 \\ 0 & 0 & 1 \end{pmatrix}$$

b. Find by inspection of the results of part (a) the transformation matrix for a twofold rotation axis lying in the x–y-plane and making an angle of θ with the x-axis.

10.2 Construct a representation of the point group C_{3v} using the transformation of coordinates method and display its decomposition into irreducible representations.

THE TRANSFORMATION OF FUNCTIONS

The rotation, reflection, inversion, or whatever of a function is really intuitively obvious after we draw a few pictures of what is occurring. The matter is often clouded when we attempt to say in mathematical terms what has been sketched in pictures. Because it is important to *use* the mathematics and because it is much faster than drawing pictures, we need to know how to handle the symbolism. A similar situation exists for vectors; once a pictorial scheme is understood, the mathematical symbolism is easier to digest.

In Figure 2 are shown two vectors $\vec{r}_1$ and $\vec{r}_2$ (a) and two function $\psi_1(\mathbf{r})$ and $\psi_2(\mathbf{r})$ (b) and (c), where for the function argument we use the column vector representation of $\vec{r}$, $\mathbf{r}$. ψ_1 and ψ_2 are taken to be of identical "shape" but oriented differently in space; they could be two atomic p-orbitals, for example. The analytical form of ψ_1 might be taken to be

$$\psi_1(\mathbf{r}) = (y + ax)e^{-(x^2+y^2)}, \qquad 0 < a < 1 \tag{10.11}$$

ψ_1 is antisymmetric under inversion and tends to a positive maximum at some r where both x and y are positive and where $y > x$.

Let us define $\hat{O}_R$ as a clockwise rotation of 90°. Clearly $\hat{O}_R$ takes the vector $\vec{r}_1$, above into $\vec{r}_2$

$$\mathbf{R}\mathbf{r}_1 = \mathbf{r}_2 \tag{10.12}$$

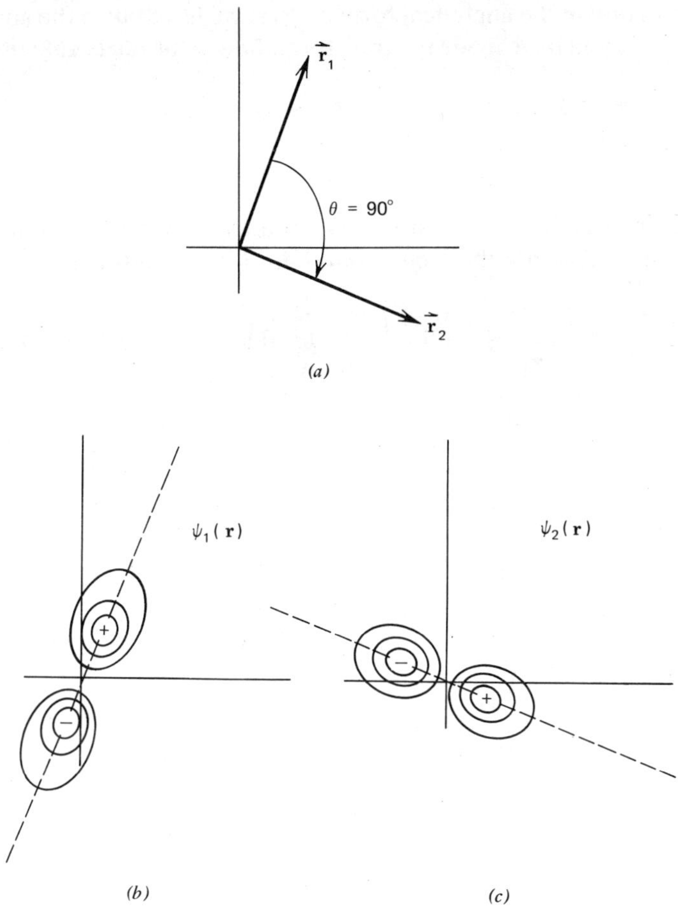

Figure 2

where **R** is the matrix representation of $\hat{O}_R$ and $\mathbf{r}_1$ and $\mathbf{r}_2$ are the column vector representations of $\vec{r}_1$ and $\vec{r}_2$. In a consistent manner let $\hat{O}_R$ rotate the function ψ_1 by 90° in a clockwise manner; that is, we rotate the contours of ψ_1 to new positions. Now the action of $\hat{O}_R$ on $\psi_1(\mathbf{r})$ has produced a new function which we call $\psi_2(\mathbf{r})$.

$$\hat{O}_R\psi_1(\mathbf{r}) = \psi_2(\mathbf{r}) \tag{10.13}$$

as should be intuitively clear from the drawing. Both ψ_1 and ψ_2 are functions of **r**; in fact, it should be easy to see that, given the analytical form of ψ_1 above, ψ_2 is given by

$$\psi_2(\mathbf{r}) = (x - ay)e^{-(x^2+y^2)} \tag{10.14}$$

In order to obtain the analytical form of $\psi_2(\mathbf{r})$ we have taken the analytical form of $\psi_1(\mathbf{r})$ and have made the transformation (coordinate substitution)

$$\begin{aligned} x &\to -y \\ y &\to +x \\ z &\to +z \end{aligned} \tag{10.15}$$

Now the matrix $\mathbf{R}$ for a positive 90° rotation according to our earlier discussion concerning the *transformation of coordinates* is

$$\mathbf{R} = \begin{pmatrix} 0 & 1 & 0 \\ -1 & 0 & 0 \\ 0 & 0 & 1 \end{pmatrix} \tag{10.16}$$

which performs the operation

$$\mathbf{R}\begin{pmatrix} x \\ y \\ z \end{pmatrix} = \begin{pmatrix} y \\ -x \\ z \end{pmatrix} \tag{10.17}$$

The operation indicated in Eq. (10.15) is, however, $\mathbf{R}^{-1}(\theta \to -\theta, +90° \to -90°)$

$$\mathbf{R}^{-1}\begin{pmatrix} x \\ y \\ z \end{pmatrix} = \begin{pmatrix} 0 & -1 & 0 \\ 1 & 0 & 0 \\ 0 & 0 & 1 \end{pmatrix}\begin{pmatrix} x \\ y \\ z \end{pmatrix} = \begin{pmatrix} -y \\ x \\ z \end{pmatrix} \tag{10.18}$$

We can, therefore, write the following equation:

$$\hat{O}_R\psi_1(\mathbf{r}) = \psi_1(\mathbf{R}^{-1}\mathbf{r}) \tag{10.19}$$

where $\psi_1(\mathbf{R}^{-1}\mathbf{r})$ is that function (here, $\psi_2(\mathbf{r})$) formed by taking the analytical form of ψ_1 and changing the variables according to the coordinate transformation $\mathbf{R}^{-1}\mathbf{r}$. Stated an equivalent way, the transformed function $\hat{O}_R\psi_1(\mathbf{r})$ $(= \psi_1(\mathbf{R}^{-1}\mathbf{r}) = \psi_2(\mathbf{r}))$ has a value at $r_0(x_0, y_0, z_0)$ that ψ_1 has at $\mathbf{R}^{-1}\mathbf{r}_0$. In order to prevent errors in manipulating the symbolism we must remember to write equations such as (10.19) with an $\mathbf{r}$ in the argument on the left-hand side and an $\mathbf{R}^{-1}\mathbf{r}$ on the other. That is, operators must only operate on functions. $\hat{O}_R\psi_1(\mathbf{r})$ and $\psi_2(\mathbf{r})$ are *different* functions and writing $\hat{O}_R\psi_1(\mathbf{r}) = \psi_2(\mathbf{r}) = \psi_1(\mathbf{R}^{-1}\mathbf{r})$ means that $\psi_2(\mathbf{r}) = \psi_1(\mathbf{R}^{-1}\mathbf{r})$ is an equality between two different functions at particular related coordinates [$\mathbf{r}$ for $\psi_2(\mathbf{r})$, and $\mathbf{R}^{-1}\mathbf{r}$ for ψ_1]. It is not proper to "operate" on both sides of such an equality; rather, we must let a subsequent operator operate on the function that the first operator has generated and *then* use the equalities between the various functions at related points in space. The subsequent discussion will further clarify this point.

SUBSTITUTION OPERATORS

In the previous illustration the operator $\hat{O}_a$ when operating on $\psi_1(\mathbf{r})$ produced the new function $\psi_1(\mathbf{A}^{-1}\mathbf{r})$, where $\mathbf{A}$ was the matrix representation of $a \in G$ and $\mathbf{A}^{-1}\mathbf{r}$ was substituted for $\mathbf{r}$. $\hat{O}_a$ acts thus as a substitution operator on functions. Given two such operations we find

$$\begin{aligned}
\hat{O}_a\phi(\mathbf{r}) &= \phi(\mathbf{A}^{-1}\mathbf{r}) = \phi'(\mathbf{r}) \\
\hat{O}_b\hat{O}_a\phi(\mathbf{r}) &= O_b\phi'(\mathbf{r}) = \phi''(\mathbf{r}) \\
&= \phi'(\mathbf{B}^{-1}\mathbf{r}) \qquad (10.20) \\
&= \phi(\mathbf{A}^{-1}\mathbf{B}^{-1}\mathbf{r}) \\
&= \phi((\mathbf{BA})^{-1}\mathbf{r})
\end{aligned}$$

where we have remembered to let operators operate only on functions such as ϕ, ϕ', and ϕ''. If the matrix representations satisfy $\mathbf{C} = \mathbf{BA}$, it should be clear that

$$\begin{aligned}
\hat{O}_c\phi(\mathbf{r}) &= \phi(\mathbf{C}^{-1}\mathbf{r}) \\
&= \phi((\mathbf{BA})^{-1}\mathbf{r}) \\
&= \hat{O}_b\hat{O}_a\phi(\mathbf{r}) \qquad (10.21)
\end{aligned}$$

so that the substitution operators $\{\hat{O}_a\}$ form a group. We can, therefore, form a representation of G by letting the various $\hat{O}_a$ act on any stable space spanned by a set of functions $\{\phi_n(\mathbf{r}), n = 1, 2, \ldots, N\}$ and determine the matrix representations of the substitution operators

$$\begin{aligned}
\hat{O}_a\phi_n(\mathbf{r}) &= \phi_n(\mathbf{A}^{-1}\mathbf{r}) \\
&= \sum_{m=1}^{n} \phi_m(\mathbf{r}) D(\hat{O}_a)_{mn} \qquad (10.22)
\end{aligned}$$

or

$$\begin{aligned}
\hat{O}_a\phi &= \phi\mathbf{D}(\hat{O}_a) \\
\phi &= (\phi_1, \phi_2, \ldots, \phi_n)
\end{aligned} \qquad (10.23)$$

where just as in the case of vectors we express the basis as the row vector ϕ of basis functions. In the following section we examine several examples of this process.

EXAMPLES OF FUNCTION TRANSFORMATIONS

Consider as our first example two $1s$ atomic orbitals located at $\pm x_0$ along the x-axis (like the hydrogen molecule). Denote the two orbitals as ϕ_1 and ϕ_2 and consider the group C_2 where the twofold axis coincides with the z-axis (see Figure 3). Clearly (using simply now the symmetry element

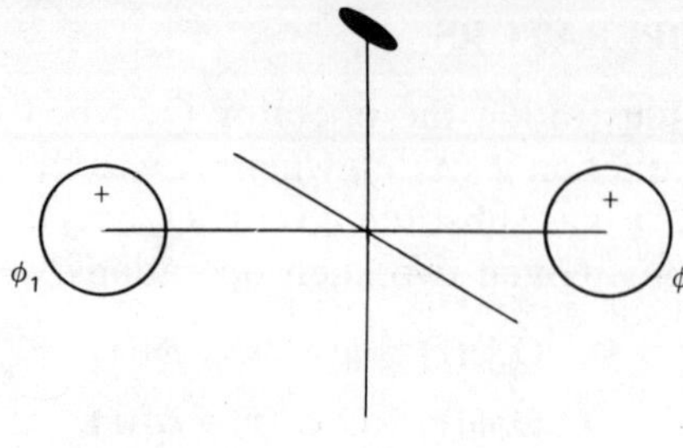

Figure 3

symbol as the operator symbol)

$$E\phi_1 = \phi_1 \qquad C_2\phi_1 = \phi_2$$
$$E\phi_2 = \phi_2 \qquad C_2\phi_2 = \phi_1 \tag{10.24}$$

so that, using the operator action on the row vector of basis functions, we get

$$\begin{aligned} E(\phi_1, \phi_2) &= (\phi_1, \phi_2)\mathbf{D}(E) \\ &= (\phi_1, \phi_2) \\ C_2(\phi_1, \phi_2) &= (\phi_1, \phi_2)\mathbf{D}(C_2) \\ &= (\phi_2, \phi_1) \end{aligned} \tag{10.25}$$

yielding

$$\mathbf{D}(E) = \begin{pmatrix} 1 & 0 \\ 0 & 1 \end{pmatrix}; \qquad \mathbf{D}(C_2) = \begin{pmatrix} 0 & 1 \\ 1 & 0 \end{pmatrix} \tag{10.26}$$

The two basis functions have led to a two-dimensional representation of C_2. Referring to the character table for C_2 we find that this representation decomposes as $\Gamma = A + B$.

As a second example, consider four 2_{pz} atomic orbitals situated at the

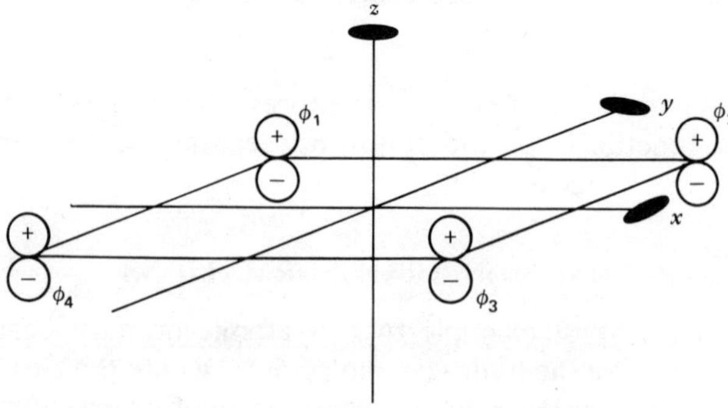

Figure 4

corners of a rectangle in the x–y-plane and consider the point group D_2. (See Figure 4.) Keeping the signs of the orbital functions in mind, we find

$$\begin{aligned}
E(\phi_1, \phi_2, \phi_3, \phi_4) &= (\phi_1, \phi_2, \phi_3, \phi_4)\\
C_2^{(z)}(\phi_1, \phi_2, \phi_3, \phi_4) &= (\phi_3, \phi_4, \phi_1, \phi_2)\\
C_2^{(y)}(\phi_1, \phi_2, \phi_3, \phi_4) &= (-\phi_2, -\phi_1, -\phi_4, -\phi_3)\\
C_2^{(x)}(\phi_1, \phi_2, \phi_3, \phi_4) &= (-\phi_4, -\phi_3, -\phi_2, -\phi_1)
\end{aligned} \tag{10.27}$$

so that

$$\mathbf{D}(E) = \begin{pmatrix} 1 & 0 & 0 & 0\\ 0 & 1 & 0 & 0\\ 0 & 0 & 1 & 0\\ 0 & 0 & 0 & 1 \end{pmatrix}; \qquad \mathbf{D}(C_2^{(z)}) = \begin{pmatrix} 0 & 0 & 1 & 0\\ 0 & 0 & 0 & 1\\ 1 & 0 & 0 & 0\\ 0 & 1 & 0 & 0 \end{pmatrix}$$

$$\mathbf{D}(C_2^{(y)}) = \begin{pmatrix} 0 & -1 & 0 & 0\\ -1 & 0 & 0 & 0\\ 0 & 0 & 0 & -1\\ 0 & 0 & -1 & 0 \end{pmatrix}; \tag{10.28}$$

$$\mathbf{D}(C_2^{(x)}) = \begin{pmatrix} 0 & 0 & 0 & -1\\ 0 & 0 & -1 & 0\\ 0 & -1 & 0 & 0\\ -1 & 0 & 0 & 0 \end{pmatrix}$$

A four-dimensional representation has been formed which decomposes in D_2 according to $\Gamma = A_1 + B_1 + B_2 + B_3$.

Consider the point group C_4 (C_4 parallel to the z-axis) and the *functions* $\phi_1 = x$, $\phi_2 = y$. Treating x and y as *functions* it is clear, for example, that $C_4 x = -y$, $C_4 y = x$; here we are transforming x and y as *functions* ($O_R \psi(\mathbf{r}) = \psi(\mathbf{R}^{-1}\mathbf{r})$ and not as coordinates of a vector. Thus,

$$\begin{aligned}
E(x, y) &= (x, y); & C_4(x, y) &= (-y, x)\\
C_2(x, y) &= (-x, -y); & C_4^3(x, y) &= (y, -x)
\end{aligned} \tag{10.29}$$

and the corresponding representation is easily written:

$$\begin{matrix} E & C_4 & C_2 & C_4^3 \\ \begin{pmatrix} 1 & 0\\ 0 & 1 \end{pmatrix} & \begin{pmatrix} 0 & 1\\ -1 & 0 \end{pmatrix} & \begin{pmatrix} -1 & 0\\ 0 & 1 \end{pmatrix} & \begin{pmatrix} 0 & -1\\ 1 & 0 \end{pmatrix} \end{matrix} \tag{10.30}$$

We could extend this idea of forming representations by considering functions consisting of products of coordinates. Thus, the six product pairs x^2, y^2, z^2, xy, xz, yz will allow us to form a six-dimensional representation, the third-order polynomials x^3, x^2, y, xyz and so forth, a higher dimensional (in this case, $d = 10$) representation. Using all the polynomials of a given order usually results in a reducible representation;

the C_{3v} example of problem 10.2 illustrates the point. For C_{3v}, x and y formed a basis for an irreducible two-dimensional representation (E) while z formed a basis for a one-dimensional one (A_1). We say that the function z "transforms like" A_1 and that the function *pair* x, y transforms like E. Because of their general importance in scientific manipulations, coordinates and pair products of coordinates—treated as functions—are normally listed in character tables showing their transformation properties. Thus, for C_{3v},

C_{3v}	E	$2C_3$	$3\sigma_v$	
A_1	1	1	1	$z,\ x^2+y^2, z^2$
A_2	1	1	-1	R_z
E	2	-1	0	$(x,y)(R_x, R_y)(x^2-y^2, xy)(xz, yz)$

the function x^2+y^2 transforms like A_1 as does z^2. The pair (x^2-y^2, xy) transforms like E as does the pair (xz, yz). That is, any operation of *this* group on x^2+y^2 gives x^2+y^2 back unchanged; the group operations cause x^2-y^2 and xy to be mixed (only with each other) such that the resulting two-dimensional matrices resulting constitute one of the two E representations.

Other entries found in most tables are the quantities denoted by R_x, R_y, and R_z. Any set of three quantities $(\alpha_x, \alpha_y, \alpha_z)$ is said to be a *polar vector* if the three components transform like the three components x, y, z of an ordinary vector. Consider the three-quantity set $(\alpha_y\beta_z-\alpha_z\beta_y, \alpha_z\beta_x-\alpha_x\beta_z, \alpha_x\beta_y-\alpha_y\beta_x)$, where the α's and β's are components of polar vectors; the cross product (vector direct product) of two polar vectors results in such a set of quantities. This latter set is called an *axial vector;* its components transform like the coordinates of an ordinary vector under rotations but, contrary to the situation for coordinates, do *not* change sign under inversion. It is easy to see, for example, that inverting two vectors $\vec{r}_1 \to -\vec{r}_1$, $\vec{r}_2 \to -\vec{r}_2$ does not change their cross product

$$(-\vec{r}_1)\times(-\vec{r}_2)=\vec{r}_1\times\vec{r}_2 \qquad (10.31)$$

The quantities R_x, R_y, and R_z refer to components of a general axial vector just as x, y, and z denote the components of a general polar vector. Singly, as pairs, or together they may form bases for—transform like—irreducible representations of the group in question. They are important to consider in the case of molecular vibrations and rotations (thus the R symbolism), where the transformation properties of angular momenta ($\vec{r}\times\vec{p}$, an axial vector) are of concern. The transformation properties of R_x, R_y, and R_z can often be determined readily for a particular group by the device of "drawing a rotation" (a curved vector) about the axis in

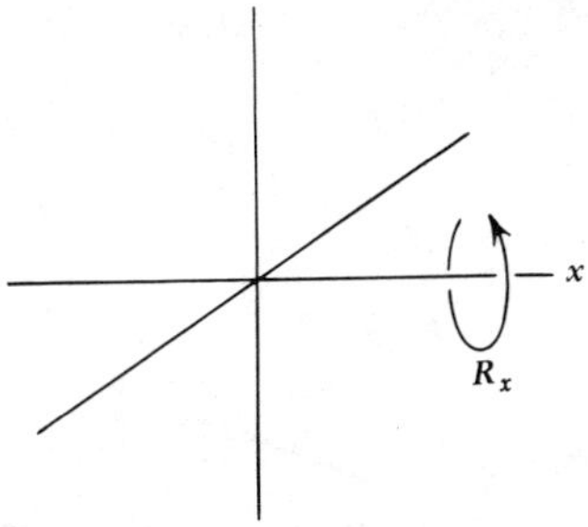

Figure 5

question and examining what happens to it under the various operations of the group (see Figure 5). Thus, in C_{3v}, R_z is unaffected by the operations E, C_3, and C_3^2 but changes sign (reverses its direction) under σ_1, σ_2, and σ_3; this clearly corresponds to the transformation properties of A_2. On the other hand, R_x and R_y are mixed by some of the operations of the group; they clearly cannot transform as a one-dimensional representation and, accordingly, must belong to E as a pair. This simple approach to the transformation properties of the components of axial vectors and its relations to properties of rotations is not accidental and stems from the very important relationship between angular momentum operators and rotation operators.

These examples serve to illustrate how functions may serve as a basis set for a representation of a group. The procedure is rather straightforward and generally leads to reducible representations. Of greater concern is finding within such sets new subsets which serve as basis sets of *irreducible* representations. We begin our attack on this problem in the next chapter.

SOLUTIONS TO PROBLEMS

10.1 a. z is clearly unchanged by reflection in a vertical plane so we concentrate on how a vector $\vec{r}$ in the x–y-plane transforms. (See Figure 6.)

$$\left.\begin{aligned} x' &= r\cos(\eta - 2\phi) \\ y' &= r\sin(\eta - 2\phi) \end{aligned}\right\} \quad \text{or} \quad \left\{\begin{aligned} \eta &= \theta + \phi \\ \phi &= \eta - \theta \end{aligned}\right.$$

so

$$x' = r\cos(2\theta - \eta) = x\cos 2\theta + y\sin 2\theta$$
$$y' = r\sin(2\theta - \eta) = x\sin 2\theta - y\cos 2\theta$$

and

$$\mathbf{D}(\sigma_v(\theta_x)) = \begin{pmatrix} \cos 2\theta & \sin 2\theta & 0 \\ \sin 2\theta & -\cos 2\theta & 0 \\ 0 & 0 & 1 \end{pmatrix}$$

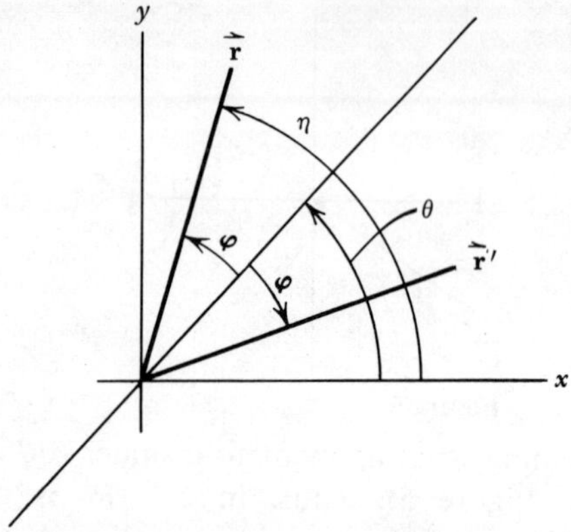

Figure 6

b. The difference between part a and this problem is that while x and y behave similarly, as though a reflection were taking place, z changes to $-z$ under the twofold rotation. Thus,

$$\mathbf{D}(C_2^{(xy)}(\theta_x)) = \begin{pmatrix} \cos 2\theta & \sin 2\theta & 0 \\ \sin 2\theta & -\cos 2\theta & 0 \\ 0 & 0 & -1 \end{pmatrix}$$

10.2 None of the operations of C_{3v} changes z. We need to consider rotations about the z-axis, with rotation angles of 0 and $\pm 120°$, and three vertical reflection planes, making angles with the x-axis of $+90(\sigma_1)$, $-30(\sigma_2)$, and $-150(\sigma_3)$ or, to make the calculation simpler, $90(\sigma_1)$, $-30(\sigma_2)$, and $+30(\sigma_3)$, since we have a choice for θ for such planes that does not at all affect the result. (See Figure 7.)

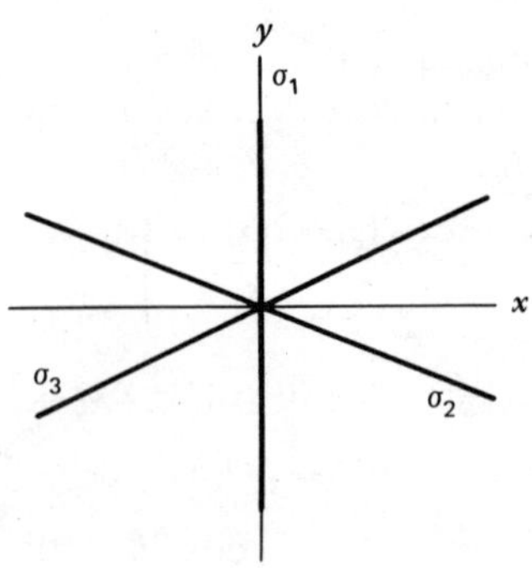

Figure 7

Accordingly,

$$E \quad \begin{pmatrix} 1 & 0 & 0 \\ 0 & 1 & 0 \\ 0 & 0 & 1 \end{pmatrix} \qquad C_3 \quad \begin{pmatrix} -\frac{1}{2} & \frac{\sqrt{3}}{2} & 0 \\ -\frac{\sqrt{3}}{2} & -\frac{1}{2} & 0 \\ 0 & 0 & 1 \end{pmatrix} \qquad C_3^2 \quad \begin{pmatrix} -\frac{1}{2} & -\frac{\sqrt{3}}{2} & 0 \\ \frac{\sqrt{3}}{2} & -\frac{1}{2} & 0 \\ 0 & 0 & 1 \end{pmatrix}$$

$$\sigma_1 \quad \begin{pmatrix} -1 & 0 & 0 \\ 0 & 1 & 0 \\ 0 & 0 & 1 \end{pmatrix} \qquad \sigma_2 \quad \begin{pmatrix} \frac{1}{2} & -\frac{\sqrt{3}}{2} & 0 \\ -\frac{\sqrt{3}}{2} & -\frac{1}{2} & 0 \\ 0 & 0 & 1 \end{pmatrix} \qquad \sigma_3 \quad \begin{pmatrix} \frac{1}{2} & \frac{\sqrt{3}}{2} & 0 \\ \frac{\sqrt{3}}{2} & -\frac{1}{2} & 0 \\ 0 & 0 & 1 \end{pmatrix}$$

The representation is clearly reducible since we recognize that it is in block diagonal form. One of the blocks corresponds to the identity representation, and, as we can easily show, the two-dimensional block belongs to the E irreducible representation. Thus, $\Gamma = A_1 + E$.

11 Projection Operators and Symmetry Adapted Functions

GENERAL PROJECTION OPERATORS

A *projection operator* projects out of an arbitrary vector or function a certain part (or parts) unchanged while annihilating (getting rid of) the rest of the vector or function. Consider ordinary three-dimensional vector space defined by the three orthonormal vectors $\vec{\varepsilon}_i$, $i = 1, 2, 3$ such that $\vec{\varepsilon}_i \cdot \vec{\varepsilon}_j = \delta_{ij}$. The "operator" $\hat{P}_1 = \vec{\varepsilon}_1\vec{\varepsilon}_1\cdot$ (note the absence of any symbol *between* the $\vec{\varepsilon}$'s specifying action but rather *after* the second $\vec{\varepsilon}$ indicating future action as an operator) projects out of a general vector that part parallel (along) $\vec{\varepsilon}_1$. Thus,

$$\begin{aligned} \vec{v} &= \alpha\vec{\varepsilon}_1 + \beta\vec{\varepsilon}_2 + \gamma\vec{\varepsilon}_3 \\ \hat{P}_1\vec{v} &= \vec{\varepsilon}_1\vec{\varepsilon}_1 \cdot \vec{v} = \alpha\vec{\varepsilon}_1\vec{\varepsilon}_1 \cdot \vec{\varepsilon}_1 + \beta\vec{\varepsilon}_1\vec{\varepsilon}_1 \cdot \vec{\varepsilon}_2 + \gamma\vec{\varepsilon}_1\vec{\varepsilon}_1 \cdot \vec{\varepsilon}_3 \\ &= \alpha\vec{\varepsilon}_1 \end{aligned} \tag{11.1}$$

where use has been made of the orthogonality properties of the $\vec{\varepsilon}_i$. Consider next an orthonormal set of functions $\{|n\rangle\}$ in Dirac notation. For an arbitrary function

$$|\psi\rangle = \sum_n c_n\,|n\rangle \tag{11.2}$$

the projection operator $\hat{P}_m = |m\rangle\langle m|$ projects out the $|m\rangle$ component destroying all the rest.

$$\begin{aligned} \hat{P}_m\,|\psi\rangle &= |m\rangle\langle m| \sum_n c_n\,|n\rangle \\ &= \sum_n |m\rangle\langle m \mid n\rangle c_n \\ &= c_m\,|m\rangle \end{aligned} \tag{11.3}$$

We can, of course, project out at the same time as many components as we wish. In the two illustrations above $\vec{\varepsilon}_1\vec{\varepsilon}_1\cdot + \vec{\varepsilon}_2\vec{\varepsilon}_2\cdot$ and $|1\rangle\langle 1| + |2\rangle\langle 2|$ will project out the sum of the "1" and "2" components in each respective case. The operator $\vec{\varepsilon}_1\vec{\varepsilon}_1\cdot + \vec{\varepsilon}_2\vec{\varepsilon}_2\cdot + \vec{\varepsilon}_3\vec{\varepsilon}_3\cdot$ projects out everything in the first case and is equivalent to the unit operator. For the orthonormal basis $\{|n\rangle\}$ the unit operator would be

$$1 = \sum_m |m\rangle\langle m| \tag{11.4}$$

Generally, given some set of basis functions (or basis vectors) $\{\phi_i\}$ we can define a set of projection operators $\{\hat{P}_i\}$ such that

$$\hat{P}_i\phi_j = \delta_{ij}\phi_j \tag{11.5}$$

It is not necessary to have an orthogonal basis to define projection operators but, as is often the case with orthogonal bases, it is usually more convenient.

Problems

11.1 Using the definition of the projection operator as given in Eq. (11.5) show that

a. $\hat{P}_i^2 = \hat{P}_i$

b. $\hat{P}_i\hat{P}_j = 0, \quad i \neq j$

c. $\sum_i \hat{P}_i = 1$

11.2 In the example of the ordinary three-dimensional vector space, find the matrix representations of $\vec{\varepsilon}_1\vec{\varepsilon}_1\cdot$, $\vec{\varepsilon}_2\vec{\varepsilon}_2\cdot$, and $\vec{\varepsilon}_3\vec{\varepsilon}_3\cdot$.

11.3 a. Consider the class of functions of a single variable x. Let $\hat{p}$ denote the operator that changes the sign of x in $f(x)$

$$\hat{p}f(x) = f(-x)$$

Show that the operator $\frac{1}{2}(1+p)$ projects out of an arbitrary function $f(x)$ that part which is symmetric in x.

b. Construct the operator that projects out the antisymmetric part of an arbitrary function $f(x)$.

GROUP PROJECTION OPERATORS

Consider a group of elements $s \in G$ and its isomorphic operator group with f: $s \to \hat{O}_s \in G$. Let the set $\{\phi_i^{(\alpha)}\}$ be a basis for the αth irreducible representation

$$\hat{O}_s\phi_i^{(\alpha)} = \sum_j \phi_j^{(\alpha)} D^{(\alpha)}(s)_{ji} \tag{11.6}$$

where $\phi_j^{(\alpha)}$ is said to "belong to" the jth row of Γ_α since, in the expansion of $\hat{O}_s\phi_i^{(\alpha)}$, $\phi_j^{(\alpha)}$ appears with $D^{(\alpha)}(s)_{ji}$, j denoting the row index of the matrix $D(s)$. The operator

$$\hat{P}_{ki}^{(\alpha)} = \frac{d_\alpha}{{}^oG} \sum_{s\in G} D^{(\alpha)}(s^{-1})_{ik}\hat{O}_s \qquad \text{(general)}$$

$$= \frac{d_\alpha}{{}^oG} \sum_{s\in G} D^{(\alpha)}(s)^*_{ki}\hat{O}_s \qquad \text{(for a unitary representation)} \tag{11.7}$$

when operating on a general $\phi_j^{(\alpha)}$ has, using the same approach employed in problem 8.5, the properties

$$\hat{P}_{ki}^{(\alpha)}\phi_j^{(\beta)} = \phi_k^{(\alpha)}\delta_{\alpha\beta}\delta_{ji} \tag{11.8}$$

That is, $\hat{P}_{ki}^{(\alpha)}$ will replace $\phi_i^{(\alpha)}$ with $\phi_k^{(\alpha)}$ and will annihilate all other basis functions. It is, in general, *not* a projection operator since

$$(\hat{P}_{ki}^{(\alpha)})^2 = 0 \neq \hat{P}_{ki}^{(\alpha)} \qquad \text{for} \qquad k \neq i \tag{11.9}$$

If, however, $k = i$, $\hat{P}_{kk}^{(\alpha)}$ does satisfy the projection operator requirements,

$$\begin{aligned} \hat{P}_{kk}^{(\alpha)}\phi_j^{(\beta)} &= \phi_k^{(\alpha)}\delta_{\alpha\beta}\delta_{kj} \\ (\hat{P}_{kk}^{(\alpha)})^2\phi_j^{(\beta)} &= (\hat{P}_{kk}^{(\alpha)})\phi_j^{(\beta)} \end{aligned} \tag{11.10}$$

The power of the "diagonal" ($k = i$) and "off-diagonal" ($k \neq i$) operators is readily seen; given an arbitrary function which can be expressed in terms of the basis sets $\{\phi_i^{(\alpha)}\}$, $\alpha = 1, 2, \ldots$

$$\phi = \sum_{i,\alpha} c_i^{(\alpha)}\phi_i^{(\alpha)} \tag{11.11}$$

$\hat{P}_{kk}^{(\beta)}$ projects out the one term $c_k^{(\beta)}\phi_k^{(\beta)}$ while, following this result, $\hat{P}_{ik}^{(\beta)}$ acting on $c_k^{(\beta)}\phi_k^{(\beta)}$ produces, for each i, $c_k^{(\beta)}\phi_i^{(\beta)}$. The coefficient $c_k^{(\beta)}$ is unimportant insofar as it is not zero; the important point is that from an arbitrary function we can create or project out the set $\{\phi_k^{(\beta)}\}$. Knowledge of the group representation Γ_β and the operation of $\hat{O}_s$ on the arbitrary function permits us then to construct a basis for the irreducible representation Γ_β.

Several remarks are in order here and are best made by more carefully specifying the types of function spaces we deal with. Given the conditions all functions of a particular problem must satisfy (the boundary conditions) in a given space, that space is spanned by a complete set of linearly independent functions f_i (a basis) if any arbitrary function satisfying the boundary conditions can be expanded in terms of the f_i

$$f = \sum_i a_i f_i \tag{11.12}$$

The set $\{f_i\}$ need have nothing to do with the basis sets for irreducible representations. If the operator group $G = \{\hat{O}_s\}$ acting on each f_i produces another function or combination of functions in the space clearly the basis $\{f_i\}$ is stable under the operations of G and can be used to form a representation of G. In fact, the problem we are concerned with is finding new basis subsets and stable subspaces to form bases for irreducible representations.

The size of the set $\{f_i\}$ may vary from problem to problem. Given a group with irreducible representations of dimension d_α, $\alpha = 1, 2, \ldots, n$ we know that there are *at least* $\sum d_\alpha$ *distinct* basis functions needed to form *irreducible* basis sets of *every* irreducible representation. If the dimension of the space spanned by $\{f_i\}$ is less than this number we are obviously not going to find—using our operator technique outlined above—basis functions for all the irreducible representations. If the dimension of our space is equal to or higher than $\sum d_\alpha$ we will generally find more than one set of functions that transform like (form a basis for) the same given representation and may be able to find at least one basis set for each irreducible representation.

Often, in order to be complete, a space must be spanned by an infinite set; such is the case in many chemical problems of molecules and atoms. In such cases, for obvious practical reasons we are forced to use approximate methods that usually involve employing a finite and incomplete basis set. The remarks made in the previous paragraph still apply.

In general, then, a given basis set—approximate or exact, incomplete or complete—may provide none or several subsets of basis functions that will carry a particular irreducible representation of a group. By systematic application of the operators $\hat{P}^{(\alpha)}_{kk}$ and $\hat{P}^{(\alpha)}_{ik}$ we will be able to decompose the original basis set into sets spanning some or all of the irreducible representations of G.

Problem

11.4 Consider the three functions $f_1 = x$, $f_2 = y$, $f_3 = z$ and use the operators $\hat{P}^{(\alpha)}_{kk}$ and $\hat{P}^{(\alpha)}_{ik}$ to project basis sets spanning the irreducible representations of C_{3v}. Use the unitary representations as given in problem 8.4.

The power of the $\hat{P}^{(\alpha)}_{ki}$ operators is that they can be used to generate functions transforming as specific rows of irreducible representations. The difficulty in using them lies in the fact that one must know and use the entire matrix of each group element in the representation. A more readily employed operator is given by summing the diagonal operators $\hat{P}^{(\alpha)}_{kk}$

$$\begin{aligned}\hat{P}^{(\alpha)} &\equiv \sum_k \hat{P}^{(\alpha)}_{kk} \\ &= \frac{d_\alpha}{{}^{\circ}G} \sum_k \sum_{s \in G} D^{(\alpha)}(s^{-1})_{kk} \hat{O}_s \\ &= \frac{d_\alpha}{{}^{\circ}G} \sum_{s \in G} \chi^{(\alpha)}(s^{-1}) \hat{O}_s \\ &= \frac{d_\alpha}{{}^{\circ}G} \sum_{s \in G} \chi^{(\alpha)}(s)^* \hat{O}_s \end{aligned} \qquad (11.13)$$

where the last equation is generally valid and follows from the equivalence of all representations to unitary representations. $\hat{P}^{(\alpha)}$ is a projection operator that will project out of an arbitrary function that part which transforms as Γ_α. We may show that $\hat{P}^{(\alpha)}$ is a projection operator by proving that

$$\hat{P}^{(\alpha)}\hat{P}^{(\beta)} = \delta_{\alpha\beta}\hat{P}^{(\alpha)} \tag{11.14}$$

This is done using the form of $\hat{P}^{(\alpha)}$, the explicit equation for the trace, and the rearrangement theorem. Thus

$$\begin{aligned}\frac{({}^{o}G)^2}{d_\alpha d_\beta}\hat{P}^{(\alpha)}\hat{P}^{(\beta)} &= \sum_{s,r} \chi^{(\alpha)}(s)^*\chi^{(\beta)}(r)^*\hat{O}_s\hat{O}_r \\ &= \sum_{s,r,k,i} D^{(\alpha)}(s^{-1})_{kk}D^{(\beta)}(r^{-1})_{ii}\hat{O}_s\hat{O}_r\end{aligned} \tag{11.15}$$

Let $sr = t$, $r^{-1} = t^{-1}s$ so $\hat{O}_s\hat{O}_r = \hat{O}_{sr} = \hat{O}_t$. Then

$$\begin{aligned}\frac{({}^{o}G)^2}{d_\alpha d_\beta}\hat{P}^{(\alpha)}\hat{P}^{(\beta)} &= \sum_{s,t,k,i} D^{(\alpha)}(s^{-1})_{kk}D^{(\beta)}(t^{-1}s)_{ii}\hat{O}_t \\ &= \sum_{s,t,k,i,j} D^{(\alpha)}(s^{-1})_{kk}D^{(\beta)}(t^{-1})_{ij}D^{(\beta)}(s)_{ji}O_t \\ &= \sum_{t,k,i,j}\left(\frac{{}^{o}G}{d_\alpha}\right)\delta_{ki}\delta_{kj}D^{(\beta)}(t^{-1})_{ij}\hat{O}_t\delta_{\alpha\beta} \\ &= \frac{{}^{o}G}{d_\alpha}\sum_{t,k} D^{(\beta)}(t^{-1})_{kk}\hat{O}_t\delta_{\alpha\beta}\end{aligned} \tag{11.16}$$

or

$$\begin{aligned}\hat{P}^{(\alpha)}\hat{P}^{(\beta)} &= \frac{d_\alpha}{{}^{o}G}\sum_{t} \chi^{(\beta)}(t)^*\hat{O}_t\delta_{\alpha\beta} \\ &= \hat{P}^{(\alpha)}\delta_{\alpha\beta}\end{aligned} \tag{11.17}$$

The above proof may seem a bit tedious, but it does serve to illustrate use of several theorems we have learned before. It could have been shown just as readily by considering the effect of $\hat{P}^{(\alpha)}$ on $\phi_j^{(\alpha)}$. Thus, since

$$\begin{aligned}\hat{P}_{ki}^{(\alpha)}\phi_j^{(\beta)} &= \delta_{\alpha\beta}\delta_{ji}\phi_k^{(\alpha)} \\ \hat{P}^{(\alpha)}\phi_j^{(\beta)} &= \sum_i P_{ii}^{(\alpha)}\phi_j^{(\beta)} \\ &= \delta_{\alpha\beta}\sum_i \phi_i^{(\beta)}\delta_{ji} = \delta_{\alpha\beta}\phi_j^{(\beta)}\end{aligned} \tag{11.18}$$

and it is easily seen that $\hat{P}^{(\alpha)}$ does indeed satisfy Eq. (11.14). The above equation also indicates what is going to happen when $\hat{P}^{(\alpha)}$ acts on a general function; if an arbitrary function contains any $\phi_j^{(\alpha)}$, all j, $\hat{P}^{(\alpha)}$ will project out that part of the function. Thus, if

$$\begin{aligned}\phi &= \sum_{j,\beta} c_j^{(\beta)}\phi_j^{(\beta)}\\ \hat{P}^{(\alpha)}\phi &= \sum_{j,\beta} c_j^{(\beta)}\hat{P}^{(\alpha)}\phi_j^{(\beta)}\\ &= \sum_{j,\beta} c_j^{(\beta)}\delta_{\alpha\beta}\phi_j^{(\beta)}\\ &= \sum_{j} c_j^{(\alpha)}\phi_j^{(\alpha)}\end{aligned} \tag{11.19}$$

The advantage of the $\hat{P}^{(\alpha)}$ projection operators is that we need to know only the characters instead of the entire matrix of the irreducible representation. A disadvantage is that, as seen in the above equation, $\hat{P}^{(\alpha)}$ produces a *linear combination* of the $\phi_j^{(\alpha)}$ of a given Γ_α. In order to determine a basis for Γ_α we must collect the results of $\hat{P}^{(\alpha)}f_i$, for each f_i of the space, and form the desired proper linearly independent sets.

Some examples will serve to illustrate the use of character projection operators. The examples we treat first are from the previous chapter where we first considered the transformation of functions and how functions could be used to form representations of groups.

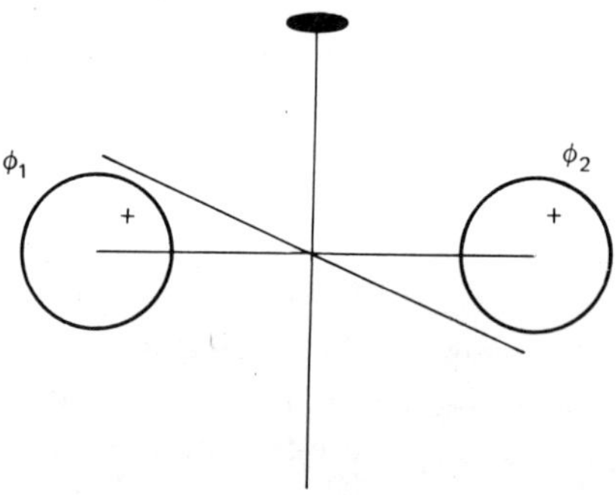

Figure 1

Example 1. The H_2-like system (see Figure 1). For simplicity the group is taken to be C_2 with the character table given below. As part of the table we include the functions involved and the results of operating on them with the various symmetry operators of the group; thus, in the row labeled by ϕ_1, the entries under the columns E and C_2 represent the action of the operators E and C_2 on ϕ_1, namely, $E\phi_1 = \phi_1$ and $C_2\phi_1 = \phi_2$.

C_2	E	C_2
A	1	1
B	1	-1
ϕ_1	ϕ_1	ϕ_2
ϕ_1	ϕ_2	ϕ_1
Γ	2	0

Finally, the last row labeled Γ indicates the character of the (generally reducible) representation formed by the basis $\{\phi_1, \phi_2\}$; the character of Γ is readily determined by observing that for each operator the trace will be given by summing the coefficients of ϕ_i in each row labeled by ϕ_i. That is, the character is the summed measure of the extent to which each of the basis functions is individually unchanged under a given operation and this is readily read off the table. The character of Γ and its decomposition tell us what irreducible representations are contained in Γ; each time an irreducible representation Γ_α occurs in Γ we *will* (certainty, not possibility) find a *set* (of dimension d_α) of functions transforming like that irreducible representation. In this first example, $\Gamma = A + B$, and we are guaranteed that the use of our character projection operator will provide us with one linearly independent function transforming as A and one as B.

Consider $\hat{P}^{(A)} = \frac{1}{2}[(1)E + (1)C_2]$. Its action on ϕ_1, and ϕ_2 is

$$\begin{aligned} \hat{P}^{(A)}\phi_1 &= \tfrac{1}{2}[E\phi_1 + C_2\phi_1] = \tfrac{1}{2}(\phi_1 + \phi_2) \\ \hat{P}^{(A)}\phi_2 &= \tfrac{1}{2}[E\phi_2 + C_2\phi_2] = \tfrac{1}{2}(\phi_2 + \phi_1) \end{aligned} \tag{11.20}$$

The reason for the form in which the table has been constructed should now be apparent in that the above equations are readily obtained by multiplying the rows for A and ϕ_1 and for A and ϕ_2. Remember that, generally, it is $\chi^{(\alpha)}(s)^*$, the complex conjugate character, that is used; in this example the characters are real and $\chi^{(\alpha)}(s)^* = \chi^{(\alpha)}(s)$. $\hat{P}^{(A)}$ acting on ϕ_1 and ϕ_2 give the same result. That is, of all the $\hat{P}^{(A)}\phi_i$ there is only one linearly independent function, $\frac{1}{2}(\phi_1 + \phi_2)$. For $\hat{P}^{(B)} = \frac{1}{2}(E - C_2)$ we find

$$\begin{aligned} \hat{P}^{(B)}\phi_1 &= \tfrac{1}{2}(E\phi_1 - C_2\phi_1) = \tfrac{1}{2}(\phi_1 - \phi_2) \\ \hat{P}^{(B)}\phi_2 &= \tfrac{1}{2}(E\phi_2 - C_2\phi_2) = \tfrac{1}{2}(\phi_2 - \phi_1) \end{aligned} \tag{11.21}$$

Again, only one linearly independent function has been found; we can use either $\frac{1}{2}(\phi_1 - \phi_2)$ or $\frac{1}{2}(\phi_2 - \phi_1)$. As the character analysis of the reducible representation showed we have obtained basis functions (one each) for the two one-dimensional representations A, B.

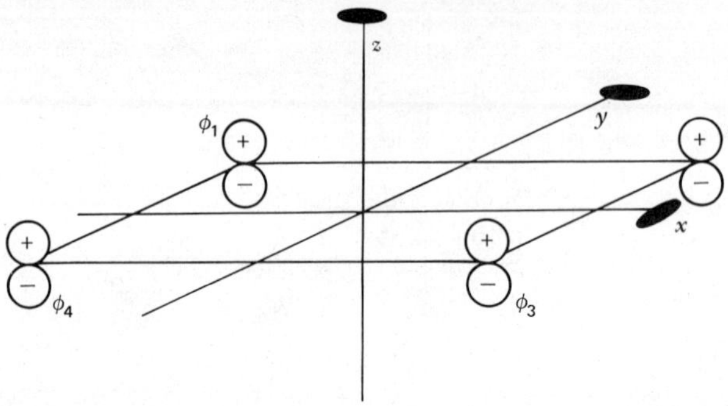

Figure 2

Example 2. Four $2p_z$ atomic orbitals at the corners of a rectangle (see Figure 2). We choose the point group D_2 (although, clearly, the implied molecular symmetry is higher) and construct the working character table as before (see Table 1).

Table 1

D_2	E	$C_2^{(z)}$	$C_2^{(y)}$	$C_2^{(x)}$
A	1	1	1	1
B_1	1	1	−1	−1
B_2	1	−1	1	−1
B_3	1	−1	−1	1
ϕ_1	ϕ_1	ϕ_3	$-\phi_2$	$-\phi_4$
ϕ_2	ϕ_2	ϕ_4	$-\phi_1$	$-\phi_3$
ϕ_3	ϕ_3	ϕ_1	$-\phi_4$	$-\phi_2$
ϕ_4	ϕ_4	ϕ_2	$-\phi_3$	$-\phi_1$
Γ	4	0	0	0
	$\Gamma = A + B_1 + B_2 + B_3$			

Consider A and $\hat{P}^{(A)}\phi_1$

$$\hat{P}^{(A)}\phi_1 = \tfrac{1}{4}(\phi_1 + \phi_3 - \phi_2 - \phi_4) \tag{11.22}$$

We can stop at this point since the decomposition of Γ has told us A appears *once*; we must find *one* function transforming like A and we have done so. Consideration of the other $\hat{P}^{(A)}\phi_i$, $i = 2, 3, 4$, will reveal that no new (linearly independent) functions are generated. In like manner we then proceed to find

$$\begin{aligned} \hat{P}^{(B_1)}\phi_1 &= \tfrac{1}{4}(\phi_1 + \phi_3 + \phi_2 + \phi_4) \\ \hat{P}^{(B_2)}\phi_1 &= \tfrac{1}{4}(\phi_1 - \phi_3 - \phi_2 + \phi_4) \\ \hat{P}^{(B_3)}\phi_1 &= \tfrac{1}{4}(\phi_1 - \phi_3 + \phi_2 - \phi_4) \end{aligned} \tag{11.23}$$

These examples have been relatively straightforward since we have found that all representations occur, and each only once. The following exercises will help provide practice and experience with situations more generally encountered.

Problems

11.5 Use the character projection operator technique to find what basis sets for irreducible representations can be generated for the following groups G and functions sets $f = \{f_i\}$.

a. $G = C_3, f = \{x, y\}$
b. $G = D_4, f = \{x^2, y^2, z^2, xy, xz, yz\}$
c. $G = C_{3v}, f = \{\phi_1, \phi_2, \phi_3\}$

where each ϕ_i represents an atomic $1s$-like orbital at the corner of an equilateral triangle.

11.6 Consider the six $2p_z$ orbitals of benzene. By inspection, write down the character of the representation of D_{6h} for which they form a basis and indicate the symmetry properties of the basis sets that can be projected out.

THE USE OF SYMMETRY ADAPTED FUNCTIONS

What we have learned to do with our group projection operators is to generate from any basis set that itself can be used to form a representation of a group subsets of functions that serve as basis sets for *irreducible* representations of the group. We have adapted our original basis to more optimally reflect the symmetry of the group; we have constructed *symmetry adapted functions*. The reason this is so important is that in quantum mechanical systems wave functions transforming like (belonging to, forming basis sets for) *different* irreducible representations of the symmetry group of the Hamiltonian have vanishing matrix elements with the Hamiltonian operator. The complexity of a given problem is usually significantly reduced by employing symmetry adapted wave functions; often symmetry considerations alone will yield the proper wave functions for a system. The basis of much of what we will do in the remainder of this book rests on our mastery of the use of group projection operators and the construction and recognition of symmetry adapted functions.

SOLUTIONS TO PROBLEMS

11.1 a. If $\hat{P}_i\phi_j = \delta_{ij}\phi_j$

$$\hat{P}_i^2\phi_j = \hat{P}_i(\delta_{ij}\phi_j)$$

$$= \delta_{ij}^2\phi_j = \delta_{ij}\phi_j = \hat{P}_i\phi_j$$

An operator $\hat{\Omega}$ such that $\hat{\Omega}^2 = \hat{\Omega}$ is said to *idempotent*. It is clear that for such an operator

$$\hat{\Omega}^n = \hat{\Omega}, \qquad n \geq 1$$

$\hat{P}_i^{-1}$, however, does not exist; thus, if

$$\hat{P}_1\phi_2 = 0$$

there is no operator that can create ϕ_2 back from 0 (the "null vector"); stated in geometrical terms, knowing the x-component of an ordinary vector does not allow one to uniquely construct the vector from which the particular x-component was projected.

b. $\hat{P}_i\hat{P}_k\phi_l = \hat{P}_i(\hat{P}_k\phi_l) = \hat{P}_i\delta_{kl}\phi_l = \delta_{il}\delta_{kl}\phi_l$

$$= 0 \quad \text{if} \quad i \neq k$$

$$= \delta_{il}\phi_l \quad \text{if} \quad i = k$$

c. Since $\hat{P}_i \sum_n c_n\phi_n = c_i\phi_i$

$$\sum_i \hat{P}_i \sum_n c_n\phi_n = \sum_i c_i\phi_i = 1 \cdot \sum_n c_n\phi_n$$

or

$$\sum_i \hat{P}_i = 1$$

11.2 In matrix terms $\hat{P}_1$ has the effect

$$\hat{P}_1\begin{pmatrix} x \\ y \\ z \end{pmatrix} = \begin{pmatrix} x \\ 0 \\ 0 \end{pmatrix}; \qquad \mathbf{P}_1 = \begin{pmatrix} 1 & 0 & 0 \\ 0 & 0 & 0 \\ 0 & 0 & 0 \end{pmatrix}$$

In like manner

$$\mathbf{P}_2 = \begin{pmatrix} 0 & 0 & 0 \\ 0 & 1 & 0 \\ 0 & 0 & 0 \end{pmatrix}; \qquad \mathbf{P}_3 = \begin{pmatrix} 0 & 0 & 0 \\ 0 & 0 & 0 \\ 0 & 0 & 1 \end{pmatrix}$$

and we see that

$$\mathbf{P}_1 + \mathbf{P}_2 + \mathbf{P}_3 = \begin{pmatrix} 1 & 0 & 0 \\ 0 & 1 & 0 \\ 0 & 0 & 1 \end{pmatrix} = \mathbf{1}$$

It is clear that the P's have no inverse since each matrix representation of P is singular (the determinants vanish).

11.3 a. Let $s(x)$ and $a(x)$ be the symmetric and antisymmetric parts, respectively, of an arbitrary function $f(x)$: $s(-x) = s(x)$ and $a(-x) = -a(x)$. Any function can be so decomposed.

$$f(x) = \underbrace{\tfrac{1}{2}(f(x) + f(-x))}_{s(x)} + \underbrace{\tfrac{1}{2}(f(x) - f(-x))}_{a(x)}$$

Then

$$\tfrac{1}{2}(1+\hat{p})f(x)=\tfrac{1}{2}f(x)+\tfrac{1}{2}pf(x)$$
$$=\tfrac{1}{2}(f(x)+f(-x))$$
$$=s(x)$$

b. The antisymmetric projecting operator when operating on $f(x)$ must yield $a(x)=\tfrac{1}{2}(f(x)-f(-x))$ and is clearly given by $\tfrac{1}{2}(1-\hat{p})$.

11.4 For convenience we reproduce the representations given in problem 8.4:

	E	σ_1	σ_2	σ_3	C_3	C_3^2
Γ_1	1	1	1	1	1	1
Γ_2	1	-1	-1	-1	1	1
Γ_3	$\begin{pmatrix}1&0\\0&1\end{pmatrix}$	$\begin{pmatrix}-1&0\\0&1\end{pmatrix}$	$\begin{pmatrix}1/2&-\sqrt{3}/2\\-\sqrt{3}/2&-1/2\end{pmatrix}$	$\begin{pmatrix}1/2&\sqrt{3}/2\\\sqrt{3}/2&-1/2\end{pmatrix}$	$\begin{pmatrix}-1/2&\sqrt{3}/2\\-\sqrt{3}/2&-1/2\end{pmatrix}$	$\begin{pmatrix}-1/2&-\sqrt{3}/2\\\sqrt{3}/2&-1/2\end{pmatrix}$

In the expressions $\hat{P}_{ki}^{(\alpha)}=(d_\alpha/{}^\circ G)\sum_{s\in G}D_{ki}^{(\alpha)}(s)^*\hat{O}_s$, we shall take the order of appearance of the various terms in the sum to be that in the above table: E first, σ_1 second, and so forth. The designation of the operations is as given in Figure 3. Note that one is transforming a *function* now and not coordinates:

$$\hat{O}_R f_i(r)=f_i(R^{-1}r)$$

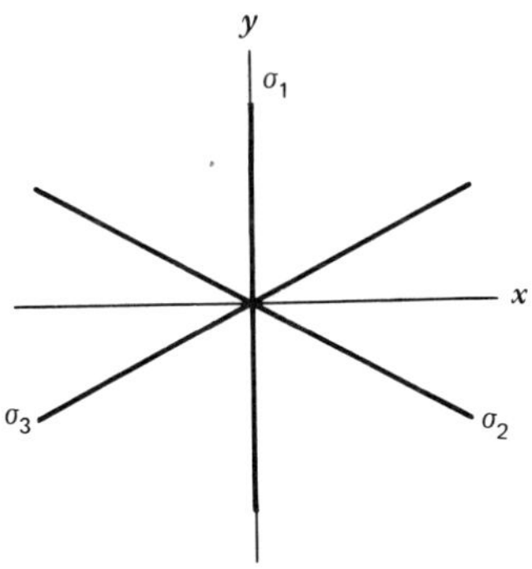

Figure 3

For Γ_1 there are only "diagonal" operators. Thus,

$$6P_{11}^{(1)}x=(1)x+(1)(-x)+(1)\left(\tfrac{1}{2}x-\frac{\sqrt{3}}{2}y\right)+(1)\left(\tfrac{1}{2}x+\frac{\sqrt{3}}{2}y\right)$$
$$+(1)\left(-\tfrac{1}{2}x-\frac{\sqrt{3}}{2}y\right)+(1)\left(-\tfrac{1}{2}x+\frac{\sqrt{3}}{2}y\right)$$
$$=0$$

$6P_{11}^{(1)}y=0$, likewise, as can be argued from the standpoint of the "equivalence" of x and y in this point group.

$$6P_{11}^{(1)}z=(1)z+(1)z+(1)z+(1)z+(1)z+(1)z=6z$$

Again in Γ_2 only diagonal operators are present here. We need not test $f_3=z$ because we have already found that z transforms like Γ_1; thus, $P_{11}^{(2)}$ can only annihilate it.

$$\begin{aligned}6P_{11}^{(2)}x &= (1)x+(-1)(-x)+(-1)\left(\tfrac{1}{2}x-\frac{\sqrt{3}}{2}y\right)+(-1)\left(\tfrac{1}{2}x+\frac{\sqrt{3}}{2}y\right)\\ &\quad +(1)\left(-\tfrac{1}{2}x-\frac{\sqrt{3}}{2}y\right)+(1)\left(-\tfrac{1}{2}x+\frac{\sqrt{3}}{2}y\right)\\ &=0\end{aligned}$$

$6P_{11}^{(2)}y=0$ also.

In the case of Γ_3 let us begin with $P_{11}^{(3)}$ on x and then operate with $P_{21}^{(3)}$ on the result. We could, of course, also perform the process by using $P_{22}^{(3)}$ and $P_{12}^{(3)}$ if we wished to.

$$\begin{aligned}\tfrac{6}{2}P_{11}^{(3)}x &= (1)x+(-1)(-x)+(\tfrac{1}{2})\left(\tfrac{1}{2}x-\frac{\sqrt{3}}{2}y\right)\\ &\quad +(\tfrac{1}{2})\left(\tfrac{1}{2}x+\frac{\sqrt{3}}{2}y\right)\\ &\quad +(-\tfrac{1}{2})\left(-\tfrac{1}{2}x-\frac{\sqrt{3}}{2}y\right)+(-\tfrac{1}{2})\left(-\tfrac{1}{2}x+\frac{\sqrt{3}}{2}y\right)\\ &=3x\end{aligned}$$

$$\begin{aligned}\tfrac{6}{2}P_{21}^{(3)}x &= (0)x+(0)(-x)+\left(-\frac{\sqrt{3}}{2}\right)\left(\tfrac{1}{2}x-\frac{\sqrt{3}}{2}y\right)\\ &\quad +\left(\frac{\sqrt{3}}{2}\right)\left(\tfrac{1}{2}x+\frac{\sqrt{3}}{2}y\right)\\ &\quad +\left(-\frac{\sqrt{3}}{2}\right)\left(-\tfrac{1}{2}x-\frac{\sqrt{3}}{2}y\right)+\left(\frac{\sqrt{3}}{2}\right)\left(-\tfrac{1}{2}x+\frac{\sqrt{3}}{2}y\right)\\ &=3y\end{aligned}$$

Thus, the *functions* $f_1=x$, $f_2=y$ as a pair transform as E while $f_3=z$ transforms as A_1 of C_{3v}.

This problem illustrates nicely the difference between transforming coordinates and transforming functions.

11.5 Remember to use $\chi^{(\alpha)}(s)^*$ (complex conjugate) and to transform the f_i as functions we proceed.

a. We treat C_3 as consisting of three one-dimensional irreducible representations.

C_3	E	C_3	C_3^2	$\varepsilon = e^{2\pi i/3}$
A	1	1	1	$= -\frac{1}{2} + i\frac{\sqrt{3}}{2}$
$E\ \Gamma_2$	1	ε^*	ε	$\frac{1}{2}(\varepsilon + \varepsilon^*) = -\frac{1}{2}$
$E\ \Gamma_3$	1	ε	ε^*	$\frac{1}{2}(\varepsilon - \varepsilon^*) = i\frac{\sqrt{3}}{2}$
x	x	$-\frac{1}{2}x - \frac{\sqrt{3}}{2}y$	$-\frac{1}{2}x + \frac{\sqrt{3}}{2}y$	
y	y	$\frac{\sqrt{3}}{2}x - \frac{1}{2}y$	$-\frac{\sqrt{3}}{2}x - \frac{1}{2}y$	
Γ	2	-1	-1	$\Gamma = \Gamma_2 + \Gamma_3$

A will not appear. For Γ_2

$$3\hat{P}^{(\Gamma_2)}x = x + (\varepsilon^*)^*\left(-\tfrac{1}{2}x - \frac{\sqrt{3}}{2}y\right) + (\varepsilon)^*\left(-\tfrac{1}{2}x + \frac{\sqrt{3}}{2}y\right)$$

$$= x(1 - \tfrac{1}{2}\varepsilon - \tfrac{1}{2}\varepsilon^*) - \frac{\sqrt{3}}{2}y(\varepsilon - \varepsilon^*)$$

$$= \tfrac{3}{2}x - i\tfrac{3}{2}y$$

$$\hat{P}^{(\Gamma_2)}x = \tfrac{1}{2}(x - iy)$$

This is the one function we need for Γ_2 and so we proceed directly to Γ_3.

$$3\hat{P}^{(\Gamma_3)}x = x + \varepsilon^*\left(-\tfrac{1}{2}x - \frac{\sqrt{3}}{2}y\right) + \varepsilon\left(-\tfrac{1}{2}x + \frac{\sqrt{3}}{2}y\right)$$

$$= x(1 - \tfrac{1}{2}\varepsilon^* - \tfrac{1}{2}\varepsilon) - \frac{\sqrt{3}}{2}y(\varepsilon^* - \varepsilon)$$

$$= \tfrac{3}{2}x + i\tfrac{3}{2}y$$

$$\hat{P}^{(\Gamma_3)}x = \tfrac{1}{2}(x + iy)$$

The fact that $\hat{P}^{(\Gamma_3)}x = (\hat{P}^{(\Gamma_2)}x)^*$ is no accident and follows directly from the fact that Γ_2 and Γ_3 are complex conjugate representations.

Had we proceeded to find, for example, $\hat{P}^{(\Gamma_3)}y$ we would have obtained

$$\hat{P}^{(\Gamma_3)}y = \tfrac{1}{2}(y - ix) = -i(\tfrac{1}{2}x + \tfrac{1}{2}iy)$$

revealing that $\hat{P}^{(\Gamma_3)}x$ and $\hat{P}^{(\Gamma_3)}y$ are not linearly independent.

b. Here is an example where classes contain more than one operator. We must explicitly express the operation of each separate element of each class while maintaining the integrity of the character table. We do so in Table 2 by separating classes by vertical lines. Although the irreducible representation basis function may be obvious from an examination of the table below, we proceed to project out the desired basis sets by brute force (see Figure 4).

Table 2

	E	$2C_4$		C_2	$2C_2'$		$2C_2''$	
D_4	E	C_4	C_4^3	C_2	$C_2'^{(1)}$	$C_2'^{(2)}$	$C_2''^{(1)}$	$C_2''^{(2)}$
A_1	1	1		1	1		1	
A_2	1	1		1	−1		−1	
B_1	1	−1		1	1		−1	
B_2	1	−1		1	−1		1	
E	2	0		−2	0		0	
x^2	x^2	y^2	y^2	x^2	x^2	x^2	y^2	y^2
y^2	y^2	x^2	x^2	y^2	y^2	y^2	x^2	x^2
z^2	z^2	z^2	z^2	z^2	z^2	z^2	z^2	z^2
xy	xy	$-xy$	$-xy$	xy	$-xy$	$-xy$	xy	xy
xz	xz	$-yz$	yz	$-xz$	$+xz$	xz	$-yz$	$-yz$
yz	yz	xz	$-xz$	$-yz$	$-yz$	$-yz$	$-xz$	$-xz$
Γ	6	0		2	2		2	

$$\Gamma = 2A_1 + B_1 + B_2 + E$$

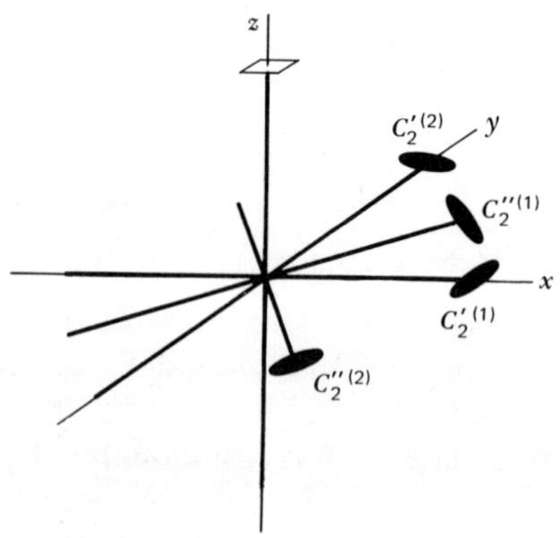

Figure 4

Since A_1 appears twice we must let $\hat{P}^{(A_1)}$ operate until *two* linearly independent basis functions are obtained. Thus,

$$\hat{P}^{(A_1)}x^2 = \tfrac{1}{2}(x^2 + y^2)$$
$$\hat{P}^{(A_1)}y^2 = \tfrac{1}{2}(x^2 + y^2)$$
$$\hat{P}^{(A_1)}z^2 = z^2$$

and at this point, having found two linearly independent A_1-type functions, we may stop. For B_1 we find

$$\hat{P}^{(B_1)}x^2 = \tfrac{1}{2}(x^2 - y^2)$$

For B_2

$$\hat{P}^{(B_2)}x^2 = 0$$
$$\hat{P}^{(B_2)}y^2 = 0$$
$$\hat{P}^{(B_2)}xy = xy$$

where we know already that $\hat{P}^{(B_2)}z^2$ would give zero. (In fact, the same conclusion could be reached for x^2 and y^2 since they individually represent mixtures of A_1 and B_1.) E appears once, and, being doubly degenerate, we must find one *pair* of linearly independent functions. Since we know that x^2, y^2, z^2, and xy are "used up" we expect xz and yz—the only two remaining functions—to transform as E. And, indeed, they do:

$$\hat{P}^{(E)}xz = xz$$
$$\hat{P}^{(E)}yz = yz$$

c. We proceed with less need for explanation. (See Figure 5.)

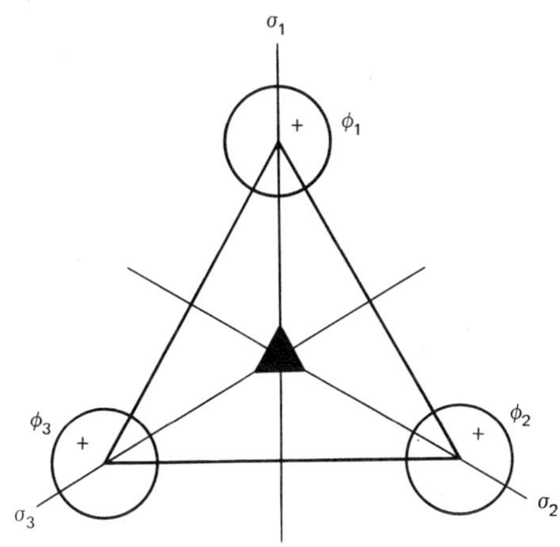

Figure 5

C_{3v}	E E	C_3	$2C_3$ C_3^2	σ_1	3σ σ_2	σ_3
A_1	1		1		1	
A_2	1		1		−1	
E	2		−1		0	
ϕ_1	ϕ_1	ϕ_2	ϕ_3	ϕ_1	ϕ_3	ϕ_2
ϕ_2	ϕ_2	ϕ_3	ϕ_1	ϕ_3	ϕ_2	ϕ_1
ϕ_3	ϕ_3	ϕ_1	ϕ_2	ϕ_2	ϕ_1	ϕ_3
Γ	3		0		1	

$$\Gamma = A_1 + E$$

$$A_1\!:\hat{P}^{(A_1)}\phi_1 = \tfrac{1}{3}(\phi_1 + \phi_2 + \phi_3)$$

$$E\!:\hat{P}^{(E)}\phi_1 = \tfrac{1}{3}(2\phi_1 - \phi_2 - \phi_3)$$

$$\hat{P}^{(E)}\phi_2 = \tfrac{1}{3}(2\phi_2 - \phi_1 - \phi_3)$$

$$\hat{P}^{(E)}\phi_3 = \tfrac{1}{3}(2\phi_3 - \phi_1 - \phi_2)$$

where we have let $\hat{P}^{(E)}$ operate on all three ϕ_i, although we need to keep only the first two produced above. The three functions produced are linearly dependent as can be seen by the fact that their sum vanishes. Our procedure has been to take the first number of linearly independent functions created; we could, however, choose any pair of the above three or any pair of any linear combination of the above three. For example, it is not uncommon to find people selecting

$$\hat{P}^{(E)}\phi_1 = \tfrac{1}{3}(2\phi_1 - \phi_2 - \phi_3)$$

$$\tfrac{1}{2}(\hat{P}^{(E)}\phi_2 - \hat{P}^{(E)}\phi_3) = \tfrac{1}{2}(\phi_2 - \phi_3)$$

This might be the choice in a quantum mechanical problem since *if* the ϕ_i are orthogonal, the latter pair is orthogonal while the pair $2\phi_1 - \phi_2 - \phi_3$ and $2\phi_2 - \phi_1 - \phi_3$ would not be.

11.6 At this point we should be able to write down the character for a given basis by inspection. The six $2p_z$ orbitals *do* form a basis for a representation of D_{6h} since the set is stable under all the group operations. Taking the C_2'-axes and the σ_v-planes to intersect orbitals the characters are: $\chi(E) = 6$, $\chi(3C_2') = -2$, $\chi(\sigma_h) = -6$, $\chi(3\sigma_v) = +2$, with all the others equal to zero. Thus, we find

$$\Gamma = B_{2g} + E_{1g} + A_{2u} + E_{2u}$$

12
Group Theory and Quantum Mechanics

QUANTUM THEORY

Quantum mechanics is concerned with obtaining solutions to the time-dependent Schrödinger equation

$$i\hbar \frac{\partial \psi}{\partial t} = \hat{\mathcal{H}} \psi \tag{12.1}$$

where ψ is the wave function (state vector) and $\hat{\mathcal{H}}$ the quantum mechanical Hamiltonian operator. The function ψ is a probability amplitude, $\psi^* \psi$ being the associated probability density. ψ can be a probability amplitude in momentum or coordinate space and generally includes spin coordinates of the various particles involved.

In quantum theory physical observables (dynamical variables) are represented by operators where possible observable values of a dynamical variable are given by the allowed eigenvalues of the associated operator. The eigenvalues a_n of an operator $\hat{a}$ are found by solving the eigenvalue problem

$$\hat{a}\phi_n = a_n \phi_n \tag{12.2}$$

where ϕ_n is the eigenfunction (eigenvector) that goes with the eigenvalue a_n. There may be several ϕ_n with the same eigenvalue in which case we speak of the eigenvalue being degenerate. Since eigenvalues are postulated to be the only possible results of a physical measurement, they must be real (noncomplex); this is guaranteed by requiring operators associated with physical observables to be *hermitian*. An operator $\hat{a}$ is hermitian if

$$(\mathbf{u}, \hat{a}\mathbf{v}) = (\hat{a}\mathbf{u}, \mathbf{v}) \tag{12.3}$$

where $\mathbf{u}$, $\mathbf{v}$ are two arbitrary vectors (functions) in the space in question and the bracket implies the usual scalar product. Of particular interest are the eigenvalues of $\hat{\mathcal{H}}$; since $\hat{\mathcal{H}}$ corresponds to the classical energy, its eigenvalues represent the allowed (observable) energies of a system

$$\hat{\mathcal{H}}\phi_n = E_n \phi_n \tag{12.4}$$

The eigenfunctions and eigenvalues of the $\hat{\mathcal{H}}$ eigenvalue equation are especially important in cases where $\hat{\mathcal{H}}$ contains no explicit time dependence. In this case the solution of Eq. (12.1) becomes

$$\psi(t)=\sum_n c_n\phi_n e^{-iE_nt/\hbar} \tag{12.5}$$

where the c_n are the constants fixed by the initial conditions and the ϕ_n are the eigenfunctions (stationary states) of $\hat{\mathcal{H}}$.

The energy index n on ϕ_n above is a quantum number indicating that ϕ_n is the nth eigenfunction of $\hat{\mathcal{H}}$. That is, if $\psi=\phi_n$, a measurement of the energy will yield precisely E_n with no uncertainty. It may be possible to label the energy eigenfunction with additional quantum numbers if it is possible for ϕ_n to be an eigenfunction of operators (dynamical variables) other than $\hat{\mathcal{H}}$. This will be so if $\hat{\mathcal{H}}$ commutes with the operator in question; if more than one operator is involved, they must not only commute with $\hat{\mathcal{H}}$ but with each other. The general theorem takes the following form:

If two operators commute there exists at least one basis in which the functions are simultaneously eigenfunctions of both operators. Conversely, if a complete basis of simultaneous eigenfunctions exists for two operators, they must commute.

It is important to remember that *all* the functions are involved in the above statement and that it is not *required* that all the eigenfunctions of one operator *must* be eigenfunctions of the other. An eigenfunction of $\hat{a}$ may accidentally be an eigenfunction of $\hat{b}$ even when $\hat{a}$ and $\hat{b}$ do not commute, but in this case it is not possible to find an *entire set* that is simultaneously diagonal in $\hat{a}$ and $\hat{b}$. The permissive aspect of the theorem is best explained by direct example for it bears on our future development. Let $\hat{a}$ and $\hat{b}$ commute, and let ϕ_n be an eigenfunction of $\hat{a}$ with eigenvalue a_n. Then

$$\hat{a}(\hat{b}\phi_n)=\hat{b}\hat{a}\phi_n=a_n(\hat{b}\phi_n) \tag{12.6}$$

and clearly the function $\hat{b}\phi_n$ is also an eigenfunction of $\hat{a}$ with the *same* eigenvalue. Now if the eigenvalue a_n is unique (nondegenerate) its eigenfunction also is unique (up to a phase factor) so that $\hat{b}\phi_n$ must be proportional to ϕ_n. In this case

$$b\phi_n=b_n\phi_n \tag{12.7}$$

where the $\hat{b}$ eigenvalue b_n is simply the proportionality constant required. Thus, when a_n is unique, ϕ_n *must* be an eigenfunction of both $\hat{a}$ and $\hat{b}$. On the other hand, if ϕ_n is one of a k-fold degenerate set of functions having the same eigenvalue, the general expression for $\hat{b}\phi_n$ will be

$$\hat{b}\phi_n = \sum_{j=1}^{k}{}' \phi_j c_{jn} \tag{12.8}$$

where the primed sum indicates summation over only the k functions having the same $\hat{a}$ eigenvalue a_n. $\hat{b}\phi_n$ is an eigenfunction of $\hat{a}$ since any linear combination of degenerate eigenfunctions of the same eigenvalue will also be an eigenfunction. However, the subset of degenerate functions *need* not be eigenfunctions of $\hat{b}$; we *can*, however, always diagonalize $\hat{b}$ in each such subset so that the new linear combinations *will* simultaneously be eigenfunctions of $\hat{a}$ and $\hat{b}$. We would finally remark that these results are generally true and are not restricted to hermitian operators.

Finding operators that commute with the Hamiltonian is particularly useful in finding the stationary states themselves. Thus, if $\{\phi_n\}$ is an eigenset of functions for a hermitian operator, $\hat{a}$ that commutes with $\hat{\mathcal{H}}$, then

$$\begin{aligned} 0 &= \langle\phi_m|\ \hat{a}\hat{\mathcal{H}} - \hat{\mathcal{H}}\hat{a}\ |\phi_n\rangle \\ &= (a_m - a_n)\langle\phi_m|\ \hat{\mathcal{H}}\ |\phi_n\rangle \end{aligned} \tag{12.9}$$

so that $\langle\phi_m|\ \hat{\mathcal{H}}\ |\phi_n\rangle$ must equal zero if $a_m \neq a_n$. This means that in diagonalizing $\hat{\mathcal{H}}$ we need only consider subsets of the ϕ_n that belong to a given eigenvalue of $\hat{a}$. The above result is also true for unitary operators, a result that is important in our consideration of group symmetry operators.

A DIGRESSION ON OPERATORS AND OPERATOR REPRESENTATIVES

Given a basis set (function set, vector set) $\{\boldsymbol{\varepsilon}_i\}$ a function (vector) $\mathbf{u} = \sum_i x_i \boldsymbol{\varepsilon}_i$ can be compactly written as

$$\mathbf{u} = \boldsymbol{\varepsilon}\mathbf{x} \tag{12.10}$$

where $\mathbf{x}$ is the column vector matrix of the components of $\mathbf{u}$ and $\boldsymbol{\varepsilon}$ is a row matrix of the basis set.

$$\boldsymbol{\varepsilon} = (\boldsymbol{\varepsilon}_1, \boldsymbol{\varepsilon}_2, \ldots, \boldsymbol{\varepsilon}_n) \tag{12.11}$$

The usual scalar product of $\mathbf{u} = \boldsymbol{\varepsilon}\mathbf{x}$ and $\mathbf{v} = \boldsymbol{\varepsilon}\mathbf{y}$ is given by

$$(\mathbf{u}, \mathbf{v}) = \mathbf{x}^{\dagger}\boldsymbol{\varepsilon}^{\dagger}\boldsymbol{\varepsilon}\mathbf{y} = \mathbf{x}^{\dagger}\mathbf{M}\mathbf{y} = (\mathbf{v}, \mathbf{u})^* \tag{12.12}$$

where $\mathbf{M}$ is the (hermitian) metric matrix introduced previously. If the basis set is orthonormal $\mathbf{M}$ is simply the unit matrix. $\boldsymbol{\varepsilon}^{\dagger}$ is the transpose conjugate of $\boldsymbol{\varepsilon}$ and is the column matrix

$$\boldsymbol{\varepsilon}^{\dagger} = \begin{pmatrix} \boldsymbol{\varepsilon}_1^* \\ \boldsymbol{\varepsilon}_2^* \\ \cdot \\ \cdot \\ \cdot \\ \boldsymbol{\varepsilon}_n^* \end{pmatrix} \tag{12.13}$$

while the adjoint matrix $\mathbf{A}^{\dagger}$ is related to $\mathbf{A}$ by

$$A_{ij}^{\dagger} = A_{ji}^{*} \tag{12.14}$$

Given a particular operator $\hat{a}$ there are two ways of associating a matrix with it. The first is to denote the action of $\hat{a}$ on the basis $\{\boldsymbol{\varepsilon}_i\}$

$$\begin{aligned} \hat{a}\boldsymbol{\varepsilon} &= \boldsymbol{\varepsilon}\mathbf{A} \\ \hat{a}\boldsymbol{\varepsilon}_i &= \sum_j \boldsymbol{\varepsilon}_j A_{ji} \end{aligned} \tag{12.15}$$

The matrix $\mathbf{A}$ is often referred to as *the* matrix representative of $\hat{a}$. We can also, however, associate with $\hat{a}$ the matrix of its scalar products in the basis $\{\vec{\varepsilon}_i\}$

$$\begin{aligned} \mathbf{A}_s &= \boldsymbol{\varepsilon}^{\dagger}\hat{a}\boldsymbol{\varepsilon} \\ (\mathbf{A}_s)_{ij} &= (\boldsymbol{\varepsilon}_i, \hat{a}\boldsymbol{\varepsilon}_j) \end{aligned} \tag{12.16}$$

The matrix $\mathbf{A}_s$ we shall call the *scalar representative* and $\mathbf{A}$ the *vector representative* in order to distinguish them. It is readily seen that they are related by the metric matrix

$$\mathbf{A}_s = \boldsymbol{\varepsilon}^{\dagger}\hat{a}\boldsymbol{\varepsilon} = \boldsymbol{\varepsilon}^{\dagger}\boldsymbol{\varepsilon}\mathbf{A} = \mathbf{MA} \tag{12.17}$$

In quantum mechanical calculations we usually work with the scalar representative while our experience with group theory representations associated with operator groups tells us that we use the vector representative of the group operators. Our purpose here is to show that properties of certain types of operators are not necessarily reflected in either matrix representation alone of the operator.

The adjoint of the operator $\hat{a}$, $\hat{a}^{(a)}$, is *defined* by

$$(\hat{a}\mathbf{u}, \mathbf{v}) = (\mathbf{u}, \hat{a}^{(a)}\mathbf{v}) \tag{12.18}$$

for any two functions in the space. The inverse operator $a^{(i)}$ is *defined* by

$$(\mathbf{u}, \hat{a}^{(i)}\hat{a}\mathbf{v}) = (\mathbf{u}, \hat{a}\hat{a}^{(i)}\mathbf{v}) = (\mathbf{u}, \mathbf{v}) \tag{12.19}$$

Consider the scalar and vector representation of $a^{(a)}$

$$\hat{a}^{(a)}\boldsymbol{\varepsilon} = \boldsymbol{\varepsilon}\mathbf{A}^{(a)} \tag{12.20}$$
$$\boldsymbol{\varepsilon}^{\dagger}\hat{a}^{(a)}\boldsymbol{\varepsilon} = \mathbf{A}_s^{(a)}$$

The equation defining the adjoint operator shows us that, for $\mathbf{u} = \boldsymbol{\varepsilon}\mathbf{x}$, $\mathbf{v} = \boldsymbol{\varepsilon}\mathbf{y}$

$$\begin{gathered}(\hat{a}\mathbf{u}, \mathbf{v}) = (\mathbf{u}, \hat{a}^{(a)}\mathbf{v}) \\ \mathbf{x}^{\dagger}\mathbf{A}^{\dagger}\boldsymbol{\varepsilon}^{\dagger}\boldsymbol{\varepsilon}\mathbf{y} = \mathbf{x}^{\dagger}\boldsymbol{\varepsilon}^{\dagger}\boldsymbol{\varepsilon}\mathbf{A}^{(a)}\mathbf{y}\end{gathered} \tag{12.21}$$

This must be true for arbitrary $\mathbf{x}$, $\mathbf{y}$ so that

$$\begin{gathered}\mathbf{A}^{\dagger}\mathbf{M} = \mathbf{M}\mathbf{A}^{(a)} \\ \mathbf{A}^{(a)} = \mathbf{M}^{-1}\mathbf{A}^{\dagger}\mathbf{M}\end{gathered} \tag{12.22}$$

Further, using the general relationship between the vector and scalar representatives we find

$$\begin{aligned}\mathbf{A}_s^{(a)} &= \mathbf{M}\mathbf{A}^{(a)} \\ &= \mathbf{M}\mathbf{M}^{-1}\mathbf{A}^{\dagger}\mathbf{M} \\ &= \mathbf{A}^{\dagger}\mathbf{M} = \mathbf{A}^{\dagger}\mathbf{M}^{\dagger} = (\mathbf{M}\mathbf{A})^{\dagger} \\ \mathbf{A}_s^{(a)} &= \mathbf{A}_s^{\dagger}\end{aligned} \tag{12.23}$$

where we have used the fact that $\mathbf{M}$ is hermitian ($\mathbf{M} = \mathbf{M}^{\dagger}$). Thus, although the scalar representative of the adjoint $\hat{a}^{(a)}$ *is* the adjoint of the scalar representative of $\hat{a}$, such is not generally so for the vector representative; only in an orthogonal basis ($\mathbf{M} = \mathbf{1}$) will $\mathbf{A}^{(a)} = \mathbf{A}^{\dagger}$. The following exercises further illustrate the point we are making.

Problems

12.1 An operator is hermitian if $(\mathbf{u}, \hat{a}\mathbf{v}) = (\hat{a}\mathbf{u}, \mathbf{v})$ in general; that is $\hat{a}^{(a)} = \hat{a}$. Show that for a hermitian operator

$$\mathbf{A} = \mathbf{M}^{-1}\mathbf{A}^{\dagger}\mathbf{M}$$
$$\mathbf{A}_s^{\dagger} = \mathbf{A}_s$$

12.2 Show that for the general case of the operator inverse $\hat{a}^{(i)}$

$$\mathbf{A}^{(i)} = \mathbf{A}^{-1}$$
$$\mathbf{A}_s^{(i)} = \mathbf{M}\mathbf{A}_s^{-1}\mathbf{M}$$

12.3 A unitary operator is one for which $(\hat{a}\mathbf{u}, \hat{a}\mathbf{v}) = (\mathbf{u}, \mathbf{v})$; that is,

$\hat{a}^{(a)}\hat{a} = \hat{a}\hat{a}^{(a)} = 1$ or $\hat{a}^{(a)} = a^{(i)}$. Show that for unitary operators

$$\mathbf{A}^{-1} = \mathbf{M}^{-1}\mathbf{A}^{\dagger}\mathbf{M}$$

$$\mathbf{A}_s^{\dagger} = \mathbf{M}\mathbf{A}_s^{-1}\mathbf{M}$$

12.4 If $\{\phi_n\}$ is an eigenbasis for $\hat{a}$ with eigenvalues a_n, show that it is also an eigenbasis for $\hat{a}^{(i)}$ with eigenvalues a_n^{-1}.

12.5 Show that the eigenvalues of unitary operators are of modulus 1

$$a_n a_n^* = 1$$

12.6 If $\{\phi_n\}$ is an eigenbasis for the unitary operator $\hat{a}$ and $[\hat{a}, \hat{\mathscr{H}}]_- = 0$, show that

$$\langle\phi_m|\ \hat{\mathscr{H}}\ |\phi_n\rangle = 0 \qquad \text{if} \qquad a_m \neq a_n$$

An essential point is contained in problem 12.3 above: The vector representative of a unitary operator need be a unitary matrix only in an orthogonal basis. It turns out also to be true that vector representatives of *antiunitary* operators (see problem 12.7) obey similar relationships and are also given by unitary matrices in an orthogonal basis. (Antiunitary operators are important in consideration of time reversal symmetry.) For finite groups one of our earlier theorems stated that every representation was equivalent to a unitary representation. This means that operator groups isomorphic to an abstract group must consist of operators that are unitary or antiunitary. That this is immensely plausible can readily be seen by considering the scalar product (integral) $(\hat{O}_R\phi, \hat{O}_R\chi)$ for any two functions, $\hat{O}_R$ being a spatial symmetry operator. $\hat{O}_R$ rotates or reflects or inverts *both* ϕ and χ and has as a net effect simply a relabeling of the variables in the integral (scalar product). The scalar product must, thus, remain unchanged

$$(\hat{O}_R\phi, \hat{O}_R\chi) = (\phi, \chi) \tag{12.24}$$

showing that $\hat{O}_R$ is indeed a unitary operator.

Problem

12.7 Whereas a linear operator $\hat{b}$ satisfies

$$\hat{b}(\alpha\phi + \beta\chi) = \alpha(\hat{b}\phi) + \beta(\hat{b}\chi)$$

an *antilinear* operator $\hat{a}$ satisfies

$$\hat{a}(\alpha\phi + \beta\chi) = \alpha^*(\hat{a}\phi) + \beta^*(\hat{a}\chi)$$

for any two functions ϕ, χ and complex constants α, β. $\hat{a}$ is said to be *antiunitary* if it is antilinear and satisfies

$$(\hat{a}\phi, \hat{a}\chi) = (\chi, \phi) = (\phi, \chi)^*$$

Show that $\hat{a}$ can always be written as

$$\hat{a} = \hat{u}\hat{K}$$

where $\hat{u}$ is a linear, unitary operator and $\hat{K}$ is the operator of complex conjugation

$$\hat{K}(\alpha\phi) = \alpha^*\phi^*$$

THE SYMMETRY GROUP OF THE HAMILTONIAN

The last section indicates that operator groups must consist of operators that are unitary or antiunitary (both of which, note, preserve the *absolute* magnitude of scalar products) and that operators commuting with the Hamiltonian operator are especially useful. We arrive, then, at the definition of a *symmetry operator* and the *symmetry group* of the Hamiltonian:

A symmetry operator for a quantum mechanical system specified by the Hamiltonian $\hat{\mathscr{H}}$ is any unitary or antiunitary operator that commutes with the Hamiltonian of that system. The set of all such symmetry operators for a given system constitutes a group and is called the *symmetry group of the Hamiltonian.*

Let $G_{\mathscr{H}} = \{\hat{O}_R\}$ be such a set of symmetry operators. To see that the set is indeed a group we note that

1. $\hat{O}_E$ the identity is obviously present.
2. Unitary or antiunitary operators have inverses that, because of the hermitian nature of $\hat{\mathscr{H}}$, also commute with $\hat{\mathscr{H}}$.

$$\begin{aligned} 0 = [\hat{\mathscr{H}}, \hat{O}_R]_- &= [\hat{\mathscr{H}}, \hat{O}_R]_-^\dagger \\ &= [\hat{O}_R^\dagger, \hat{\mathscr{H}}^\dagger]_- \\ &= [\hat{O}_R^{-1}, \hat{\mathscr{H}}]_- \end{aligned}$$

3. The associative law is obeyed.
4. The product of any two operators (both unitary or both antiunitary or one of each) is also unitary or antiunitary and also commutes with $\hat{\mathscr{H}}$.

It may be worthwhile at this point to expand a bit on the physical implications of an operator commuting with the Hamiltonian operator and to discuss the various types of symmetry present in most systems. Many of the examples in this book deal with the symmetries of electronic wave functions attached to a framework of rigid (stationary) nuclei; that is, a coordinate system is defined by the nuclear framework, this nuclear framework dictating the symmetry of the system. An operation that takes any given electron from one position to another where its potential energy is exactly the same would be an appropriate symmetry operator for the system. Because the kinetic energy terms are unaffected by any point symmetry operation the symmetry of such a system is then determined by the symmetry of its potential energy. Clearly, an operator that moves an electron from one region to an equivalent region so far as a potential energy is concerned would commute with the Hamiltonian. Note how the Hamiltonian dictates the symmetry; in the present example it is the symmetry of the nuclear framework which, of course, appears as part of the Hamiltonian.

This example points up another important aspect of symmetry considerations. This aspect is the fact that we actually rotate, reflect, or invert the *relative* coordinates of particles and thereby need a frame of reference. For example, the Hamiltonian of a water molecule in free space is completely unaffected by any and all bodily rotations of the molecule. For that matter the molecule may be reflected in any mirror plane passing through the center of mass of the molecule, and this would not change the Hamiltonian. Inversion of the coordinates of all particles through the center of mass leaves the Hamiltonian unaffected whether the molecule under consideration has a center of symmetry or not! (Much of this type of discussion is expanded on in a very readable article by H. C. Longuet-Higgins.[1])

Given a frame of reference the usable symmetry elements are still very numerous for most systems. In a molecule, for example, permutation of the positions and spins of all electrons is a valid symmetry operation; indeed, the permutation of the labels of all identical particles constitute a set of valid symmetry operations. We see, thus, that a given group is actually made up of a multi-direct product of many groups dealing with different classifications of the symmetry present. The permutation symmetry of electrons (which is discussed in Chapter 14 at length) is taken into account by using a properly antisymmetrized electron wave function. Solving the electronic problem in a system of fixed nuclei gives rise to symmetry elements dictated by the nuclear configuration. In the Born–Oppenheimer approximation of the separation of electronic and nuclear motion the electronic energy expression serves as part of the potential energy for the motion of the nuclei. What is normally done is to consider the symmetry appropriate to that particular part or parts of the problem in question, that is, the symmetry determined by the partial Hamiltonian used. As the full Hamiltonian is then built up step by step the symmetry is compounded until one arrives at the full, multi-direct-product description.

SYMMETRY AND ENERGY EIGENFUNCTIONS

Let $\{\phi_n^{(j)}, n = 1, 2, \ldots, k\}$ belong to a k-fold degenerate level of $\hat{\mathscr{H}}$; that is,

$$\hat{\mathscr{H}}\phi_n^{(j)} = \lambda_j\phi_n^{(j)} \qquad n = 1, 2, \ldots, k \tag{12.25}$$

where λ_j is the eigenvalue associated with the k degenerate eigenfunctions. Since $\hat{O}_R$ and $\hat{\mathscr{H}}$ commute

$$\hat{\mathscr{H}}(\hat{O}_R\phi_n^{(j)}) = \hat{O}_R\hat{\mathscr{H}}\phi_n^{(j)} = \lambda(\hat{O}_R\phi_n^{(j)}) \tag{12.26}$$

so that the function $\hat{O}_R\phi_n^{(j)}$ must also be an eigenfunction of $\hat{\mathscr{H}}$ with the *same* eigenvalue λ_j. Generally, then, $\hat{O}_R\phi_n^{(j)}$ must be some linear combination of the k degenerate functions

$$\hat{O}_R\phi_n^{(j)} = \sum_{m=1}^{k} \phi_m^{(j)} D_{mn}(R) \tag{12.27}$$

We can easily show, as we did before in introducing the general concept of representations, that if $\hat{O}_S\hat{O}_R = \hat{O}_{SR}$ the matrix equality $\mathbf{D}(S)\mathbf{D}(R) = \mathbf{D}(SR)$ obtains, the set $\Gamma = \{\mathbf{D}(R)\}$ is a representation of $G_{\mathscr{H}}$ and the set $\{\phi_n^{(j)}\}$ forms a basis for the representation of $G_{\mathscr{H}}$ of dimension $d = k$. We are not yet able to comment on whether Γ is a reducible representation or an irreducible one. In either event the usefulness of group theory in quantum mechanics is beginning to emerge.

If Γ for the level above is reducible we may reduce it and thereby find new linear combinations of the $\phi_n^{(j)}$ which span stable subspaces and carry irreducible representations. If Γ is irreducible the set $\{\phi_n^{(j)}\}$ constitutes an irreducible basis set. Each such level λ_j can be so treated, which means that the totality of functions $\{\phi_n^{(j)}\}$ can be separated into subsets that provide bases for irreducible representations of $G_{\mathscr{H}}$. The importance of this fact lies in the fact that two functions transforming as different irreducible representations of $G_{\mathscr{H}}$ are not coupled by the Hamiltonian. Thus, let $\chi_n^{(\alpha)}$ and $\chi_m^{(\beta)}$ be two symmetry adapted functions—*not necessarily eigenfunctions of* $\hat{\mathscr{H}}$—that transform as the nonequivalent unitary irreducible representations Γ_α and Γ_β, respectively. Recall that we can always make Γ_α and Γ_β unitary by using orthogonal bases. Consider the scalar product (integral) $(\chi_n^{(\alpha)}, \chi_m^{(\beta)})$ and assume the group symmetry operators to be unitary (as opposed to antiunitary or mixed), then

$$\begin{aligned}(\chi_n^{(\alpha)}, \hat{\mathscr{H}}\chi_m^{(\beta)}) &= (\hat{O}_R\chi_n^{(\alpha)}, \hat{O}_R\hat{\mathscr{H}}\chi_m^{(\beta)}) \\ &= (\hat{O}_R\chi_n^{(\alpha)}, \hat{\mathscr{H}}\hat{O}_R\chi_m^{(\beta)})\end{aligned} \tag{12.28}$$

since $\hat{O}_R$ and $\hat{\mathscr{H}}$ commute. The equation is true for *any* symmetry operator $\hat{O}_R$ and is, therefore, true for their averaged sum

$$(\chi_n^{(\alpha)}, \hat{\mathscr{H}}\chi_m^{(\beta)}) = \frac{1}{{}^oG}\sum_R (\hat{O}_R\chi_n^{(\alpha)}, \hat{\mathscr{H}}\hat{O}_R\chi_m^{(\beta)}) \qquad (12.29)$$

Expanding the terms on the right-hand side we get

$$\begin{aligned}(\chi_n^{(\alpha)}, \hat{\mathscr{H}}\chi_m^{(\beta)}) &= \frac{1}{{}^oG}\sum_R\sum_{n'm'} (\chi_{n'}^{(\alpha)}, \hat{\mathscr{H}}\chi_{m'}^{(\beta)}) D_{n'n}^{(\alpha)}(R)^* D_{m'm}^{(\beta)}(R) \\ &= \sum_{n'm'} (\chi_{n'}^{(\alpha)}, \hat{\mathscr{H}}\chi_{m'}^{(\beta)}) \frac{1}{{}^oG}\sum_R D_{n'n}^{(\alpha)}(R)^* D_{m'm}^{(\beta)}(R) \\ &= \sum_{n'm'} (\chi_{n'}^{(\alpha)}, \hat{\mathscr{H}}\chi_{m'}^{(\beta)}) \frac{1}{d_\alpha}\,\delta_{\alpha\beta}\delta_{n'm'}\delta_{nm} \end{aligned} \qquad (12.30)$$

The fact that the integral will vanish if $n \neq m$ is due to our use of basis functions for irreducible representations; note, too, that the above result shown is that $(\chi_n^{(\alpha)}, \hat{\mathscr{H}}\chi_n^{(\alpha)})$ is independent of n. More important, though, is the fact we set out to prove, namely that

$$(\chi_n^{(\alpha)}, \hat{\mathscr{H}}\chi_m^{(\beta)}) = 0, \qquad \Gamma_\alpha \neq \Gamma_\beta \qquad (12.31)$$

The fact that the $\chi_n^{(\alpha)}$ need not be eigenfunctions of $\hat{\mathscr{H}}$ is most noteworthy. It tells us that by symmetry adapting an arbitrary basis set (by use of our group projection operators) we need only worry about matrix elements of $\hat{\mathscr{H}}$ between functions that transform according to the same irreducible representation. That is, the matrix element for $\hat{\mathscr{H}}$ between any linear combination of $\{\chi_n^{(\alpha)}\}$ and any linear combination of $\{\chi_m^{(\beta)}\}$ will vanish for $\alpha \neq \beta$.

As we shortly disclose, eigenfunctions of $\hat{\mathscr{H}}$ *will* transform as *irreducible* representations of $G_{\mathscr{H}}$. Finding these eigenfunctions will be assisted by beginning with a symmetry adapted basis set. Resolution of the reducible–irreducible question alluded to above provides us with an even stronger tool, as we show in the following section.

It should also be clear that the above calculation is valid for $\hat{\mathscr{H}} = 1$; that is, any two functions that transform according to different irreducible representations must be orthogonal.

Problem

12.8 Consider the hypothetical 1-electron system to be described in terms of four atomic 1*s* orbitals, three located at the vertices of an equilateral triangle (ϕ_1, ϕ_2, ϕ_3) and one (ϕ_4) located at the triangle's geometrical center and above the plane of the other three (a trigonal pyramid arrangement). Construct symmetry adapted 1-electron functions and demonstrate that functions transforming as different irreducible representations are not connected by the

Hamiltonian. The symmetry of the Hamiltonian is presumed to be C_{3v}.

THE IRREDUCIBILITY POSTULATE

As has been shown in the previous section, the eigenfunctions associated with a particular eigenvalue that is n-fold degenerate form the basis for an n-dimensional representation Γ of the symmetry group of the Hamiltonian. If the level is unique—nondegenerate—then the resulting representation is of dimension one and is clearly irreducible. If, however, n is greater than one it does *not* automatically follow that one has an irreducible representation of dimension n. Jansen and Boon[2] comment that on a purely group theoretical basis we cannot draw any further conclusion concerning the possible irreducibility of the representation.

In order to hopefully clarify the postulatory nature of our end result we reproduce some of the basic arguments that others have employed in arriving at the conclusion that such Γ's are irreducible. Eyring, Walter, and Kimball[3], for example, argue in this manner: If Γ were reducible the eigenvalues associated with the stable subspaces "could" be different, the fact that they are not indicating irreducibility. It does not seem possible, however, to rule out stable subspaces in this manner and Jansen and Boon indicate we cannot do so rigorously.

Other authors take a more pragmatic point of view. Since, as a matter of fact, "normally" such representations *are* irreducible and only "accidentally" does one find such a representation to be reducible—referred to as *accidental* or *nonessential* degeneracy—they consider that it is a rule seldom violated. Referring to accidental degeneracy Wigner[4] says: "It will be assumed that this is a very uncommon situation, and that it occurs in the important case of the Schrödinger equation only exceptionally." Hamermesh[5] comments: "In most cases this representation will be irreducible." And, continuing, "So we may assume, in general, that the set of eigenfunctions belonging to a given energy are a set of partners, and form the basis for one of the irreducible representations of the symmetry groups." Meijer and Bauer[6] give a very clear presentation of the relationship of irreducible representations and the Hamiltonian eigenvectors and, although they do not specifically mention the irreducibility postulate, seem to demonstrate its general validity. In short, since it is found to be true nearly all the time its general validity is postulated. Thus, the *irreducibility postulate:*

Eigenfunctions of the Hamiltonian corresponding to an n-fold degenerate energy level form a basis for an irreducible representation of $G_{\mathcal{H}}$ of dimension n.

The implications of the irreducibility postulate are rather significant and bear emphasizing. Once the appropriate symmetry group for a given system (defined by its Hamiltonian) has been determined, the *possible* degeneracies to the energy eigenvalues have been ascertained. Clearly, if each n-fold set of degenerate eigenfunctions provides a basis for an irreducible representation of dimension n, that irreducible representation of dimension n *must* be present in the symmetry group representations for that system. Note, however, that the postulate indicates only *possible* degeneracy. Thus, if a quantum mechanical system is characterized by C_{3v} symmetry, it is not *required* that one will find singly and doubly degenerate energy levels but rather that singly and doubly degenerate levels *may* occur and—barring "accidental" degeneracies—no others.

The postulate is useful whether one is attempting an exact or an approximate solution. Given any basis, approximate (truncated, incomplete) or exact (complete), we proceed to form a transformed basis set of the same order (same number of functions) of symmetry adapted functions by employing, for example, our character projection operators. The basis will divide into subsets of functions that transform as particular irreducible representations of the symmetry group in question. Every time Γ_α of dimension d_α occurs the irreducible postulate guarantees us that an energy level (approximate or exact, depending upon the completeness of our basis) corresponding to the degeneracy d_α will be found. Thus, since our original (total) basis forms a basis for a generally reducible representation of the group, a decomposition of this reducible representation by our standard procedures immediately tells us the types and numbers of degeneracy to expect in our problem. The irreducibility postulate connects the mathematical properties of the abstract group representations—their dimensionality—to a physically realizable property—the possible degeneracy of energy levels.

Problem

12.9 Use the irreducibility postulate to predict the essential (minimum required) degeneracies to be expected in the model problems given in problem 11.5 assuming that the functions there presented are some approximate sets of 1-electron functions for the solution of a 1-electron Hamiltonian.

ACCIDENTAL DEGENERACY

The irreducibility postulate says that most degeneracies are essential (determined by the symmetry group of the Hamiltonian) and that only occasionally will nonessential or accidental degeneracies be found. Essential and nonessential are good descriptive terms; group theory with the irreducibility postulate tells us of the essential (must be possible) cases of degeneracy but does not rule out others (nonessential) which may just "happen accidentally." For example, it may happen that for a particular choice of parameters in a problem that two levels of differing symmetry may cross; that is, for a particular set of circumstances they are degenerate. Such would be a nonessential or accidental degeneracy. To cite perhaps a trivial example but one which, nevertheless, illustrates the point, consider the two 1-electron molecular orbitals for a model calculation of the hydrogen molecule where the two orbitals are the sum and difference of two atomic $1s$ orbitals centered at the sites of the two protons. The two MO's are of A and B type symmetry in the point group C_2 (twofold axis taken perpendicular to the internuclear axis). They transform as different one-dimensional representations of the group and are, at all finite distances of separation, nondegenerate. At an infinite distance of separation, however, they obviously are degenerate. The symmetry of the system has not changed but an "accidental" degeneracy has occurred under special circumstances. The preference for the terms essential and nonessential is also illustrated by this example: The degeneracy seems hardly "accidental" to us, but rather is more aptly described as a nonessential one.

It is important to realize that the irreducibility postulate and its implications with regard to essential degeneracies rely on the use of the correct symmetry group of the Hamiltonian. This may seem to be an obvious requirement but its importance lies in the fact that often we choose to employ a group of lower symmetry than the true group, and sometimes we mistakenly (an intellectual accident!) presume a symmetry that turns out to be lower than that nature is really employing. In either case what may then appear to us to be a nonessential degeneracy may really be no accident at all but rather a reflection of the essential degeneracy required by the higher symmetry group. We can take heart in the fact that such errors—intentional or not—will not lead to incorrect results but will only have made our work harder in that we did not take full advantage of all the symmetry present. If an n-dimensional irreducible representation is present in our lower symmetry group—that is, a subgroup of the proper, higher symmetry group—that degree of degeneracy at least will be present when the higher symmetry is employed. Decrease of symmetry can lower the degree of degeneracy, but degeneracy lowering will never occur by increasing the symmetry.

We have mentioned before the very interesting symmetry of time inversion, and why it is that certain clearly one-dimensional representations in point group tables are labeled as though they are, for example, E types, implying a twofold degeneracy. We can comment on this a bit further by considering a situation where we neglect spin effects and are presented with a real (noncomplex) Hamiltonian. For example, consider such circumstances for a molecule that has precisely C_3 point group *spatial* symmetry. We know that all the irreducible representations of C_3 are one-dimensional, yet when a calculation is performed we will find those eigenfunctions transforming as the complex representations to be degenerate. An accident? No, for we have left out a symmetry operator that increases the symmetry of the system. The Hamiltonian being real should let us see that the operator of complex conjugation $\hat{K}$ will commute with it. $\hat{K}$ is antiunitary and along with the spatial symmetry operators will constitute the correct group of the Hamiltonian of higher symmetry. In the higher symmetry group the previously "accidental" or nonessential degeneracy will emerge as an essential one. It is a general rule that if the Hamiltonian is real, irreducible representations that contain complex characters will be degenerate and the predicted degeneracy in the lower symmetry group must then be doubled. Groups containing both unitary and antiunitary operators are normally not treated in elementary books; they involve the concept due to Wigner[4] of *corepresentations* in which some of our ideas concerning the manipulation of representations must be modified. By so noting in the spatial group character tables those representations whose degeneracy will be doubled when time reversal is a valid symmetry operator we should be prevented from ascribing the term accidental to those essential degeneracies required by time reversal symmetry.

Problems

12.10 The third atomic hydrogen-like orbitals in real form transform as the functions z^2, xz, yz, x^2-y^2, and xy. Use the point group tables where these functions are listed to show the predicted 1-electron degeneracies for systems exhibiting the progressively higher symmetries of D_2, D_4, and D_{4h}.

12.11 Consider a system exhibiting C_3 spatial symmetry and one that also has a Hamiltonian that commutes with $\hat{K}$, the operator of complex conjugation. Since $\hat{K}$ commutes with all the C_3 point group operations we may form the nonunitary direct-product group $C_3 \times K$. Using the functions $x+iy$ and $x-iy$ form a set of 2×2 matrices—one for each element of the direct-product group—that exhibit the transformation properties of $x+iy$ and $x-iy$ under the various operations of the group. Show that the matrix set so formed is not a true representation because the homomorphism condition $f(ab)=f(a)f(b)$ is not generally obeyed.

CONCLUSION

The purpose of this chapter has been to show the essential (nonaccidental!) connection between group theory and quantum mechanics. Thus, the n-fold degenerate eigenfunctions of a quantum mechanical system form a basis for an n-dimensional irreducible representation of the operator group consisting of all those unitary and/or antiunitary operators that commute with the Hamiltonian. From this statement stem the myriad of applications of group theoretical techniques to systems of physical interest. Our discussion of the statement's implications has centered about exact or approximate solutions to the Schrödinger equation, properties they must possess, and how we can best go about making their finding simpler. There are, however, still further important implications that follow from our basic statements and in the following chapters some of these will be developed.

References

1. H. C. Longuet-Higgins, *Mol. Phys.* **6,** 445 (1963).
2. L. Jansen and M. Boon, *Theory of Finite Groups Applications in Physics*, Wiley, New York, 1967, pp. 211–213.
3. H. Eyring, J. Walter, and G. E. Kimball, *Quantum Chemistry*, Wiley, New York, 1944, p. 186.
4. E. P. Wigner, *Group Theory*, Academic, New York, 1959, p. 119.
5. M. Hamermesh, *Group Theory*, Addison-Wesley, Reading, Mass., 1962, p. 161.
6. P. H. E. Meijer and E. Bauer, *Group Theory*, North-Holland, Amsterdam, 1962, pp. 110–120.

SOLUTIONS TO PROBLEMS

12.1 Since $\hat{a}^{(a)} = \hat{a}$ here, then

$$\mathbf{MA}^{(a)} = \mathbf{MA}$$

$$\mathbf{A}^{(a)} = \mathbf{A}$$

In general $\mathbf{A}_s^{(a)} = \mathbf{A}_s^{\dagger}$, so

$$\mathbf{A}_s^{(a)} = \mathbf{MA}^{(a)} = \mathbf{MA} = \mathbf{A}_s^{\dagger} = (\mathbf{MA})^{\dagger} = \mathbf{A}^{\dagger}\mathbf{M}^{\dagger}$$

or, since $\mathbf{M}^{\dagger} = \mathbf{M}$,

$$\mathbf{A} = \mathbf{M}^{-1}\mathbf{A}^{\dagger}\mathbf{M}$$

and $\mathbf{A}_s^{\dagger} = \mathbf{MA} = \mathbf{A}_s$.

12.2 $(\mathbf{u}, \hat{a}^{(i)}\hat{a}\mathbf{v}) = (\mathbf{u}, \hat{a}\hat{a}^{(i)}\mathbf{v}) = (\mathbf{u}, \mathbf{v})$ clearly implies $\hat{a}^{(i)}\hat{a} = \hat{a}\hat{a}^{(i)} = 1$. Thus,

$$(\mathbf{u}, \hat{a}^{(i)}\hat{a}\mathbf{v}) = \mathbf{x}^{\dagger}\boldsymbol{\varepsilon}^{\dagger}\boldsymbol{\varepsilon}\mathbf{A}^{(i)}\mathbf{A}\mathbf{y} = \mathbf{x}^{\dagger}\boldsymbol{\varepsilon}^{\dagger}\boldsymbol{\varepsilon}\mathbf{y}$$

leads to

$$\mathbf{M}\mathbf{A}^{(i)}\mathbf{A} = \mathbf{M}$$
$$\mathbf{A}^{(i)} = \mathbf{A}^{-1}$$

Also, then,

$$\begin{aligned}\mathbf{A}_s^{(i)} &= \mathbf{M}\mathbf{A}^{(i)} = \mathbf{M}\mathbf{A}^{-1} \\ &= \mathbf{M}\mathbf{A}^{-1}\mathbf{M}^{-1}\mathbf{M} \\ &= \mathbf{M}(\mathbf{M}\mathbf{A})^{-1}\mathbf{M} \\ &= \mathbf{M}\mathbf{A}_s^{-1}\mathbf{M}\end{aligned}$$

12.3 $\hat{a}^{(a)} = \hat{a}^{(i)}$ implies $\mathbf{M}\mathbf{A}^{(a)} = \mathbf{M}\mathbf{A}^{(i)}$. Since $\mathbf{A}^{(a)} = \mathbf{M}^{-1}\mathbf{A}^{\dagger}\mathbf{M}$ and $\mathbf{A}^{(i)} = \mathbf{A}^{-1}$ in general we have

$$\mathbf{A}^{-1} = \mathbf{M}^{-1}\mathbf{A}^{\dagger}\mathbf{M}$$

Furthermore, $\mathbf{A}_s^{(a)} = \mathbf{A}_s^{(i)}$ is also implied by the initial equality so that since $\mathbf{A}_s^{(a)} = \mathbf{A}_s^{\dagger}$ in general

$$\mathbf{A}_s^{(i)} = \mathbf{A}_s^{(a)} = \mathbf{A}_s^{\dagger} = \mathbf{M}\mathbf{A}_s^{-1}\mathbf{M}$$

12.4 $\hat{a}\phi_n = a_n\phi_n$

$$\hat{a}^{(i)}\hat{a}\phi_n = \phi_n = a_n\hat{a}^{(i)}\phi_n$$

so that

$$\hat{a}^{(i)}\phi_n = a_n^{-1}\phi_n$$

12.5 For a unitary operator $\hat{a}^{(a)} = \hat{a}^{(i)}$ let $\{\phi_n\}$ be an eigenbasis for $\hat{a}$. We have

$$(\phi_n, \hat{a}\phi_n) = a_n$$

and by the general properties of $\hat{a}^{(a)}$ and the results of problem 12.4 above

$$\begin{aligned}(\phi_n, \hat{a}\phi_n) &= (\hat{a}^{(a)}\phi_n, \phi_n) \\ &= (\phi_n, \hat{a}^{(a)}\phi_n)^* \\ &= (\phi_n, \hat{a}^{-1}\phi_n)^* \\ &= (a_n^{-1})^*\end{aligned}$$

Thus $a_n = (a_n^{-1})^*$ or $a_n a_n^* = 1$. Note that the eigenvalues of unitary operators are not required to be real.

12.6 The vanishing of the commutator leads to

$$\begin{aligned}0 &= \langle\phi_m|\, \hat{a}\hat{\mathscr{H}} - \hat{\mathscr{H}}\hat{a}\, |\phi_n\rangle \\ &= \langle a^{(a)}\phi_m|\, \hat{\mathscr{H}}\, |\phi_n\rangle - \langle\phi_m|\, \hat{\mathscr{H}}\, |\hat{a}\phi_n\rangle\end{aligned}$$

so that using the result of problem 12.5 we have

$$0 = ((a_m^{-1})^* - a_n)\langle\phi_m|\,\hat{\mathcal{H}}\,|\phi_n\rangle$$
$$= (a_m - a_n)\langle\phi_m|\,\hat{\mathcal{H}}\,|\phi_n\rangle$$

Thus, just as in the case of hermitian operators,

$$\langle\phi_m|\,\hat{\mathcal{H}}\,|\phi_n\rangle = 0 \qquad \text{if} \qquad a_m \neq a_n.$$

12.7 $\hat{a}$ will be antilinear since

$$\hat{a}(\alpha\phi) = \hat{u}\hat{K}(\alpha\phi) = \hat{u}(\alpha^*\phi^*)$$
$$= \hat{u}(\alpha^*\hat{K}\phi)$$
$$= \alpha^*(\hat{u}\hat{K}\phi)$$

It satisfies

$$(\hat{a}\phi, \hat{a}\chi) = (\phi, \chi)^*$$

since

$$(\hat{u}\hat{K}\phi, \hat{u}\hat{K}\chi) = (\hat{u}\phi^*, \hat{u}\chi^*)$$
$$= (\phi^*, \hat{u}^{(a)}\hat{u}\chi^*)$$
$$= (\phi^*, \chi^*)$$
$$= (\phi, \chi)^*$$

Note that no adjoint exists for $\hat{K}$ and that $\hat{K}^{(i)} = \hat{K}$

$$\hat{K}\hat{K}(\alpha\phi) = \hat{K}(\alpha^*\phi^*) = (\alpha^*\phi^*)^* = \alpha\phi$$

Note also, that since $\hat{u} = 1$ is a valid unitary operator, $\hat{K}$ itself is antiunitary.

12.8 The appropriate point group is C_{3v} and we proceed as before to construct symmetry adapted functions. (See Figure 1.)

C_{3v}	E	$2C_3$		$3\sigma_v$		
	E	C_3	C_3^2	σ_{v1}	σ_{v2}	σ_{v3}
A_1	1	1		1		
A_2	1	1		−1		
E	2	−1		0		
ϕ_1	ϕ_1	ϕ_2	ϕ_3	ϕ_1	ϕ_3	ϕ_2
ϕ_2	ϕ_2	ϕ_3	ϕ_1	ϕ_3	ϕ_2	ϕ_1
ϕ_3	ϕ_3	ϕ_1	ϕ_2	ϕ_2	ϕ_1	ϕ_3
ϕ_4	ϕ_4	ϕ_4	ϕ_4	ϕ_4	ϕ_4	ϕ_4
Γ	4	1		2		

$$\Gamma = 2A_1 + E$$

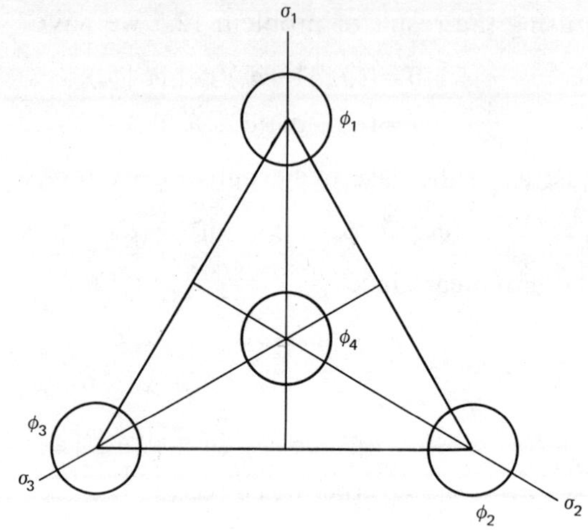

Figure 1

Symmetry adapted functions (via character projection operators)

$$A_1\text{: } \tfrac{1}{3}(\phi_1+\phi_2+\phi_3),\ \phi_4$$

$$E\text{: } \tfrac{1}{6}(2\phi_1-\phi_2-\phi_3),\ \tfrac{1}{2}(\phi_2-\phi_3)$$

Let $\mathscr{H}_{ij}=\langle\phi_i|\ \hat{\mathscr{H}}\ |\phi_j\rangle$ and calculate as an example

$$\langle\phi_1+\phi_2+\phi_3|\ \hat{\mathscr{H}}\ |\phi_2-\phi_3\rangle$$

$$\langle\phi_1+\phi_2+\phi_3|\ \hat{\mathscr{H}}\ |\phi_2-\phi_3\rangle=\mathscr{H}_{12}-\mathscr{H}_{13}+\mathscr{H}_{22}-\mathscr{H}_{23}+\mathscr{H}_{32}-\mathscr{H}_{33}$$

Because of the geometry (symmetry) $\mathscr{H}_{12}=\mathscr{H}_{13}=\mathscr{H}_{23}$ and $\mathscr{H}_{22}=\mathscr{H}_{33}=\mathscr{H}_{11}$ so the integral vanishes. On the other hand, $\langle\phi_1+\phi_2+\phi_3|\ \hat{\mathscr{H}}\ |\phi_4\rangle$ *need* not vanish since both functions transform as the same irreducible representation; indeed, in this case it does not, giving

$$\langle\phi_1+\phi_2+\phi_3|\ \hat{\mathscr{H}}\ |\phi_4\rangle=3\mathscr{H}_{14}$$

Note that we have not bothered to normalize the functions; we can always do so, and, in any event, it does not affect the orthogonality conditions.

12.9 a. $G=C_3,\ f=\{x, y\}$

$$\Gamma=\Gamma_2+\Gamma_3, \text{ the complex representations}$$

Both are one-dimensional, and—provided C_3 *is* the precise group of the Hamiltonian—no essential degeneracies are expected.

b. $G = D_4$, $f = \{x^2, y^2, z^2, xy, xz, yz\}$

$$\Gamma = 2A_1 + B_1 + B_2 + E$$

We expect four singly-degenerate levels (two A_1's, B_1, B_2) and one doubly degenerate level (E). Although A_1 occurs twice the two eigenfunctions that transform as A_1 are *not* expected to be degenerate. If E had appeared twice we would have predicted two doubly degenerate levels, not one fourfold degenerate level.

c. $G = C_{3v}$, $f = \{\phi_1, \phi_2, \phi_3\}$

$$\Gamma = A_1 + E$$

One singly degenerate and one doubly degenerate level are expected.

12.10 We copy the functions and their irreducible representations for the three groups.

	D_2	D_4	D_{4h}
z^2	A	A_1	A_{1g}
$x^2 - y^2$	A	B_1	B_{1g}
xy	B_1	B_2	B_{2g}
xz	B_2	$\{E$	$\{E_g$
yz	B_3		

For D_2 none of the functions are predicted to give rise to degeneracy. For D_4, (xz, yz) will occur as a degenerate pair and this is also true for D_{4h}. If our system had the true symmetry D_{4h} and we had used the group D_2, we would have found xz and yz to be "accidentally" degenerate. Had we used D_4 instead of the proper D_{4h} we would correctly predict the degeneracies even though we had erroneously employed the lower symmetry group. Sometimes the nonaccidental degeneracies are predicted accidentally!

12.11 For C_3 we know from an earlier problem (11.4) that $x + iy$ and $x - iy$ transform separately as the two complex one-dimensional representations of C_3. However, neither vector represents a stable subset when $\hat{K}$ is a symmetry operator since

$$K(x + iy) = x - iy$$

$$K(x - iy) = x + iy$$

The functions x and y are unaffected by $\hat{K}$, but they are mixed with one another by the C_3 operations. The best we can do, then, is form a stable two-dimensional space using $(x+iy, x-iy)$. The transformation matrices for the E, C_3, and C_3^2 operations are readily written down from our knowledge of the transformational properties of $x \pm iy$ in the group C_3

$$\mathbf{D}(E)=\begin{pmatrix}1 & 0\\ 0 & 1\end{pmatrix};\quad \mathbf{D}(C_3)=\begin{pmatrix}\varepsilon & 0\\ 0 & \varepsilon^*\end{pmatrix};\quad \mathbf{D}(C_3^2)=\begin{pmatrix}\varepsilon^* & 0\\ 0 & \varepsilon\end{pmatrix}$$

For the product operations $\hat{K}\hat{E}$, $\hat{K}\hat{C}_3$, and $\hat{K}\hat{C}_3^2$ we note that in general for a unitary operator $\hat{u}$

$$\hat{u}\phi_k = \sum_j \phi_j D_{jk}(u)$$

so that

$$\hat{K}\hat{u}\phi_K = \sum_j D_{jk}^*(u)\hat{K}\phi_j$$

$$= \sum_{jm} \phi_m D(K)_{mj} D_{jk}^*(u)$$

or

$$\mathbf{D}(Ku)=\mathbf{D}(K)\mathbf{D}(u)^*$$

which does not obey the usual homomorphism condition. Note that by a similar reasoning it will be true that

$$\mathbf{D}(uK)=\mathbf{D}(u)\mathbf{D}(K)$$

Now since $\hat{K}(x \pm iy)=(x \mp iy)$

$$\mathbf{D}(K)=\begin{pmatrix}0 & 1\\ 1 & 0\end{pmatrix}$$

It follows, then, that

$$\mathbf{D}(KC_3)=\mathbf{D}(K)\mathbf{D}(C_3)^*=\begin{pmatrix}0 & 1\\ 1 & 0\end{pmatrix}\begin{pmatrix}\varepsilon^* & 0\\ 0 & \varepsilon\end{pmatrix}$$

$$=\begin{pmatrix}0 & \varepsilon\\ \varepsilon^* & 0\end{pmatrix}$$

or, since $\hat{K}\hat{C}_3=\hat{C}_3\hat{K}$

$$\mathbf{D}(KC_3)=\mathbf{D}(C_3K)=\mathbf{D}(C_3)\mathbf{D}(K)=\begin{pmatrix}\varepsilon & 0\\ 0 & \varepsilon^*\end{pmatrix}\begin{pmatrix}0 & 1\\ 1 & 0\end{pmatrix}$$

$$=\begin{pmatrix}0 & \varepsilon\\ \varepsilon^* & 0\end{pmatrix}$$

Similarly we find

$$\mathbf{D}(KC_3^2)=\mathbf{D}(C_3^2K)=\begin{pmatrix}0 & \varepsilon^*\\ \varepsilon & 0\end{pmatrix}$$

What we have found is not a representation since the homomorphism condition is not generally obeyed, but rather is an example of what Wigner[3] calls a *corepresentation.*

The corepresentation above is irreducible, and we see that although $x+iy$ and $x-iy$ transform as different one-dimensional representations in the group C_3, for the group $C_3 \times K$ they transform as a degenerate pair in a two-dimensional irreducible corepresentation. This is why x, y are denoted as belonging to an E-type representation in C_3; one must remember that this implication will be valid only when $\hat{K}$ (or the appropriate time-reversal operator) is a valid symmetry operator.

13 Matrix Elements and Symmetry Considerations

The power of group theory applied to quantum mechanics reveals itself most clearly in the ability to predict the vanishing of matrix elements that arise in manipulations of the theory. We saw earlier that two functions transforming as different irreducible representations are orthogonal; that is, their scalar product is zero. This is just another way of saying that, given a total basis containing the two functions in question, their matrix element in the matrix representation of the unit operator is zero. Given the symmetry group of the Hamiltonian we also saw how the group projection operators allowed us to form symmetry adapted functions such that those functions transforming as different irreducible representations have zero matrix elements with the Hamiltonian operator.

In this chapter we shall demonstrate the general criteria for predicting the vanishing of general integrals or matrix elements and then proceed to illustrate the applicability of the method with several examples. We shall treat a fairly general type of integral first, show the extension to the more general type, and then look at some less general but rather important types in greater detail.

THE GENERAL CASE

Assume that a set of functions has been symmetry adapted for a given group (the group of the Hamiltonian) so that each function may be written as $\phi_{n,i}^{(\alpha)}$ where $\phi_{n,i}^{(\alpha)}$ is the ith function in the nth subset that transforms as the αth irreducible representation Γ_α. We need the extra subscript n since several *distinct* sets of functions may be present that all transform according to Γ_α. For example, n may distinguish sets of Γ_α-transforming functions that belong to different energy eigenvalues of the Hamiltonian.

Assume further that the "physical" operators (i.e., dynamical variables) of interest may also be written as $\hat{V}_j^{(\alpha)}$, implying that they also transform as particular irreducible representations. Thus, in quantum theory employing a coordinate-space representation of operators ($\hat{x} = x$, $\hat{p}_x = -ih(\partial/\partial x)$, etc.) the various operators (excluding for the moment spin operators) are functions of coordinates and as such will transform as particular irreducible representations or as combinations of particular irreducible representations. We shall assume operators transforming as a single Γ_α; the extension to the more general case of a sum of differently transforming terms should be apparent.

We want to treat matrix elements such as

$$(\phi_{n,i}^{(\alpha)}, \hat{V}_j^{(\eta)}\phi_{m,k}^{(\beta)}) \equiv \int \phi_{n,i}^{(\alpha)*}\hat{V}_j^{(\eta)}\phi_{m,k}^{(\beta)}\,dV$$

restricting our considerations to groups G consisting of unitary operators. All ordinary spatial operators are of this type, whereas time inversion, as an example, is an antiunitary operator. We must restrict ourselves to unitary groups since the theory of representations we have developed is based on that premise. As mentioned before, we must delve into the theory of corepresentations when antiunitary operators are involved.

Thus, since a scalar product is invariant to operating on both factors with a unitary operator

$$(u, v) = (\hat{O}_R u, \hat{O}_R v) \tag{13.1}$$

it will be true that for the group of operators $s \rightarrow \hat{O}_s \in G$

$$\begin{aligned}(\phi_{n,i}^{(\alpha)}, \hat{V}_j^{(\eta)}\phi_{m,k}^{(\beta)}) &= (\hat{O}_s\phi_{n,i}^{(\alpha)}, (\hat{O}_s\hat{V}_j^{(\eta)})\hat{O}_s\phi_{m,k}^{(\beta)})\\ &= \frac{1}{{}^oG}\sum_{s\in G}(\hat{O}_s\phi_{n,i}^{(\alpha)}, (\hat{O}_s\hat{V}_j^{(\eta)})\hat{O}_s\phi_{m,k}^{(\beta)})\\ &= \frac{1}{{}^oG}\sum_{s\in G}\sum_{i',j',k'}(\phi_{n,i'}^{(\alpha)}, V_{j'}^{(\eta)}\phi_{m,k'}^{(\beta)}\cdot D^{(\alpha)*}(s)_{i'i}D^{(\eta)}(s)_{j'j}D^{(\beta)}(s)_{k'k}\end{aligned} \tag{13.2}$$

Here we have used the fact that the scalar product is not only invariant to any one $\hat{O}_s$ but to their sum over $s \in G$, and that $\hat{O}_s$ operating on a "product" $(\hat{V}_j^{(\eta)}\phi_{m,k}^{(\beta)})$ transforms both terms (by definition). The resulting final form reminds us that the integrand functions ($\phi_{n,i}^{(\alpha)*}$, $\phi_{m,k}^{(\beta)}$, and the operator function $\hat{V}_j^{(\eta)}$) as a product set will normally—there are exceptions to be discussed later—form a basis for the direct-product representation Γ of G where

$$\Gamma = \Gamma_\alpha^* \times \Gamma_\eta \times \Gamma_\beta \tag{13.3}$$

As long as $n \neq m$, the separate functions $\phi_{n,i}^{(\alpha)}$ and $\phi_{m,k}^{(\beta)}$ will be independent, the resulting dimensionality of $\Gamma_\alpha^* \times \Gamma_\eta \times \Gamma_\beta$ will be $d_\alpha \cdot d_\eta \cdot d_\beta$, and the matrices of Γ will be given by the Kronecker product of the separate matrices, $\mathbf{D}(s) = \mathbf{D}^{(\alpha)}(s)^* \otimes \mathbf{D}^{(\eta)}(s) \otimes \mathbf{D}^{(\beta)}(s)$. When $n = m$ and $\alpha = \beta$ the $d_\alpha \cdot d_\eta \cdot d_\beta$ products are *not* all independent, and we must be more careful in our analysis of the direct product. We deal with this aspect a bit later.

In general Γ will be reducible so that we may write

$$\mathbf{D}(s) = \mathbf{U}^{-1}\left(\sum_\delta \oplus c_\delta \mathbf{D}^{(\delta)}(s)\right)\mathbf{U} \tag{13.4}$$

where **U** is the transformation that relates **D**(s) and its block diagonal form and c_δ represents the number of times $\mathbf{D}^{(\delta)}(s)$ appears. **D**(s) will then have the form

$$\mathbf{D}(s)=\mathbf{U}^{-1}\begin{pmatrix} \mathbf{D}^{(\varepsilon)}(s) & & & 0 \\ & \mathbf{D}^{(\varepsilon')}(s) & & \\ & & \mathbf{D}^{(\varepsilon'')}(s) & \\ & & & \ddots \\ 0 & & & \end{pmatrix}\mathbf{U} \tag{13.5}$$

Instead of writing **D**(s) as a multiple-index triple direct product we simply note the form it must have so that the term appearing on the right of (13.2) involving the sum of all $s \in G$ becomes

$$\sum_{s\in G} D(s)_{ij} = \sum_{\substack{s\in G,\\ k,m}} U_{ik}^{-1}\left(\sum_{\delta} \oplus\, c_\delta \mathbf{D}^{(\delta)}(s)\right)_{km} U_{mj}$$

$$=\sum_{k,m} U_{ik}^{-1}U_{mj}\sum_{s\in G}\left(\sum_{\delta}\oplus\, c_\delta \mathbf{D}^{(\delta)}(s)\right)_{km} \tag{13.6}$$

where we isolate the $s \in G$ sum on the right in order to invoke a special form of the grand orthogonality relations.

First, just as problem 9.8 showed us a "sum rule" for characters based on the unique character of Γ_1, the completely symmetric irreducible representation, we may perform the same trick on the grand orthogonality relations, Eqs. (8.17) and (8.18). The result is independent of whether we are using unitary representations or not, so we use (8.17), the most general form:

$$\sum_{s\in G} D^{(\alpha)}(s)_{ik} D^{(\beta)}(s^{-1})_{jl} = \left(\frac{{}^oG}{d\alpha}\right)\delta_{\alpha\beta}\delta_{il}\delta_{kj} \tag{13.7}$$

Choose $\beta = 1$, the completely symmetric one-dimensional irreducible representation where $j = l = 1$ only and $\mathbf{D}^{(1)}(s^{-1}) = 1$, all $s^{-1} \in G$. Equation (8.17) then reduces to

$$\frac{1}{{}^oG}\sum_{s\in G} D^{(\alpha)}(s)_{ik} = \delta_{\alpha 1}\delta_{i1}\delta_{k1} \tag{13.8}$$

yielding the sum rule on the individual matrix elements of the various irreducible representations of G.

The application of (13.8) to (13.6) should be clear. On the right-hand side of 13.6 the kmth element of the direct sum of the $\mathbf{D}^{(\delta)}(s)$ is either in a region where the block zeros occur—so that the kmth element is zero—or it is in a region occupied by one of the $\mathbf{D}^{(\delta)}(s)$—in which case the sum rule derived above causes it to vanish if $\delta \neq 1$. In short, *if* Γ_1 *does not appear in* $\Gamma_\alpha^* \times \Gamma_\eta \times \Gamma_\beta$ *the integral form (13.2) vanishes.*

The result derived above can be generalized to any number of functions. The result is that

$$(\phi_{n,i}^{(\alpha)} \phi_{m,j}^{(\beta)}, \ldots, \hat{V}_k^{(\eta)} \phi_{p,k}^{(\alpha')} \phi_{q,l}^{(\beta')}, \ldots)$$

will vanish unless

$$\Gamma_\alpha^* \times \Gamma_\beta^* \times \cdots \times \Gamma_\eta \times \Gamma_{\alpha'} \times \Gamma_{\beta'} \times \cdots$$

contains Γ_1 at least once. That is, all such integrals must contain a part that transforms completely symmetrically under the group in question.

DECOMPOSITION OF THE DIRECT PRODUCT; INDEPENDENT FUNCTIONS

Assume for the moment that all functions involved are independent of one another. To determine if an integral of the type under consideration has a part transforming as Γ_1 we simply decompose the direct product using our character relations.

$$\Gamma = \Gamma_\alpha \times \Gamma_\beta \times \cdots \times \Gamma_\delta = \sum_\gamma c_\gamma \Gamma_\gamma$$

$$\chi(s) = \chi^{(\alpha)}(s) \chi^{(\beta)}(s) \cdots \chi^{(\delta)}(s) = \sum_\gamma c_\gamma \chi^{(\gamma)}(s) \tag{13.9}$$

$$c_\gamma = \frac{1}{{}^oG} \sum_{s \in G} \chi^{(\gamma)}(s)^* \cdot \chi(s)$$

$$= \frac{1}{{}^oG} \sum_{s \in G} \chi^{(\gamma)}(s)^* \cdot \chi^{(\alpha)}(s) \chi^{(\beta)}(s) \cdots \chi^{(\delta)}(s)$$

where, in particular,

$$c_1 = \frac{1}{{}^oG} \sum_{s \in G} \chi^{(\alpha)}(s) \chi^{(\beta)}(s) \cdots \tag{13.10}$$

since $\chi^{(1)}(s)^* = 1$, all s.

Problems

13.1 Consider all possible integrals of the form

$$(\phi_{n,i}^{(\alpha)}, \hat{V} \phi_{m,j}^{(\beta)})$$

for $\hat{V} = x$, y, and z and where the $\phi_{n,i}^{(\alpha)}$ and $\phi_{m,j}^{(\beta)}$ are arbitrary functions transforming as Γ_α and Γ_β in the group C_{3v}. Denote those integrals that vanish according to the analysis just given.

13.2 We usually show that

$$\int_{-\infty}^{+\infty} f_e(x) f_0(x)\, dx = 0$$

where $f_e(x)$ and $f_o(x)$ are even and odd functions of x, respectively, by nongroup theoretical means, although often the statement is made that "it is zero 'by symmetry.'" Describe the vanishing of such an integral in group theoretical language.

13.3 Let $\{\phi_{n,i}^{(\alpha)}\}$ be symmetry adapted 1-electron functions for a system whose spatial symmetry is D_{6h} (for example, benzene). Indicate whether or not the 2-electron coulomb integral

$$\int \phi^{(B_{2g})}(1)^* \phi^{(B_{1g})}(2)^* \frac{1}{r_{12}} \phi^{(E_{1u})}(1) \phi^{(A_{2g})}(2)\, d\tau_1\, d\tau_2$$

vanishes or not. The operator r_{12}^{-1} transforms as A_{1g} in D_{6h}.

We can go a bit further and derive somewhat stricter integral-vanishing rules than we have done, but we shall not do so for the general case. We need to remember, however, that the above result tells us when a matrix element *must* be zero; it does *not* tell us when an integral (matrix element) must be nonzero. The presence of Γ_1 in a direct product is no guarantee of finiteness of the integral involved. For example, we would not predict the integral of two even functions to vanish "by symmetry," but if the two do not overlap anywhere their integral obviously is zero.

A SPECIFIC SITUATION

We shall illustrate a more general integral vanishing rule for a special case, namely the case where $\hat{O}_s$ and $\hat{V}$ commute in a simple two-function–one-operator situation. Although it is true (as used before) that for symmetry operators

$$\hat{O}_s(\hat{V}\phi_k^{(\beta)}) = (\hat{O}_s\hat{V})(\hat{O}_s\phi_k^{(\beta)}) \tag{13.11}$$

where $(\hat{O}_s\hat{V})$ infers $\hat{V}$ modified by $\hat{O}_s$ operating on it, it is also true by general operator methods that

$$\hat{O}_s\hat{V} = [\hat{O}_s, \hat{V}]_- + \hat{V}\hat{O}_s \tag{13.12}$$

an operator equality, where $[\hat{O}_s, \hat{V}]_-$ is the usual operator commutator. Accordingly,

$$\begin{aligned}(\phi_i^{(\alpha)}, \hat{V}\phi_j^{(\beta)}) &= (\hat{O}_s\phi_i^{(\alpha)}, \hat{O}_s\hat{V}\phi_j^{(\beta)}) \\ &= (\hat{O}_s\phi_i^{(\alpha)}, [\hat{O}_s, \hat{V}]_-\phi_j^{(\beta)} + \hat{V}(\hat{O}_s\phi_j^{(\beta)}))\end{aligned} \tag{13.13}$$

If $[\hat{O}_s, \hat{V}] = 0$ (that is, if $\hat{V}$ is unchanged by the action specified by $\hat{O}_s$) then following the lines outlined earlier we can write

$$(\phi_i^{(\alpha)}, \hat{V}\phi_j^{(\beta)}) = \frac{1}{{}^oG}\sum_{s\in G}(\phi_{i'}^{(\alpha)}, \hat{V}\phi_{j'}^{(\beta)})D^{(\alpha)}(s)^*_{i'i}D^{(\beta)}(s)_{j'j}$$

$$\propto \frac{1}{d_\alpha}\delta_{\alpha\beta}\delta_{i'j'}\delta_{ij} \qquad (13.14)$$

for unitary representations. Thus, not only if $\alpha \neq \beta$ but also if $i \neq j$, the integral vanishes. That is, even if both functions transform as the same irreducible representation, their matrix element with the commuting $\hat{V}$ operator vanishes unless they belong to the same row i of the representation. This type of integral does not often arise in the more basic applications of group theory in that we often do not have specific information about to what row a particular function belongs but, rather, simply the type of irreducible representation to which it belongs.

The most obvious examples of commuting $\hat{V}$ operators are the Hamiltonian operator (which defines the symmetry operators and the symmetry group in terms of those operators that commute with it) and the perhaps trivial but nonetheless important case of the unit operator (which commutes with everything). The importance of the first example has been given a whole previous chapter and the latter case has been employed to demonstrate the orthogonality of functions transforming as different irreducible representations.

Problems

13.4 Let $\hat{V}$ be an operator that commutes with all $\hat{O}_s \in G$. Equation (13.14) indicates that such an integral vanishes unless $\alpha = \beta$. Show that this is so since $\Gamma_\alpha^* \times \Gamma_\beta$ contains Γ_1 only if $\Gamma_\alpha = \Gamma_\beta$ (*not* $\Gamma_\alpha^* = \Gamma_\beta$).

13.5 In the general treatment of the two-function–one-operator integral involving independent functions it was shown that the matrix element was zero unless $\Gamma_\alpha^* \times \Gamma_\eta \times \Gamma_\beta$ contained Γ_1. Show that this is equivalent to demanding that $\Gamma_\eta \times \Gamma_\beta$ contain Γ_α.

13.6 For *real* representations show that $\Gamma_\alpha \times \Gamma_\eta \times \Gamma_\beta$ contains Γ_1 if $\Gamma_\alpha \times \Gamma_\beta$ contains Γ_η.

THE CASE OF DEPENDENT PRODUCT FUNCTIONS

We have stressed from time to time in our discussion to this point that we have assumed independent functions in the analysis of the direct products we have treated. It is still true in general that the general matrix element

$$(\phi_{n,i}^{(\alpha)}, \hat{V}_j^{(\eta)}\phi_{m,k}^{(\beta)})$$

will vanish unless $\Gamma_\alpha^* \times \Gamma_\eta \times \Gamma_\beta$ contains Γ_1. The situation we have avoided up to now is how to decompose such a multiple direct product when $\alpha = \beta$ *and* $n = m$, that is, the case when the product functions are not all linearly independent. Consider, for example, the partner set x, y that transforms as E in C_{3v}. Since $E \times E = A_1 + A_2 + E$, we might carelessly think that the product set $x \cdot x$, $x \cdot y$, $y \cdot x$, and $y \cdot y$ should form a basis for a four-dimensional representation of C_{3v} which would reduce to $A_1 + A_2 + E$. However, clearly among the product functions the only independent ones are $x \cdot x = x^2$, $x \cdot y = y \cdot x = xy$, and $y \cdot y = y^2$ so that only a three-dimensional representation is possible. Thus, when for our general integral above $n = m$ and $\alpha = \beta$ not all the representations $\Gamma_\alpha \times \Gamma_\alpha$ will be present; only those representations constructed from the *linearly independent* product functions $\phi_{n,i}^{(\alpha)*} \phi_{n,j}^{(\alpha)}$ will occur. The irreducible representations contained are given by the symmetrized direct product discussed below.

THE SYMMETRIZED AND ANTISYMMETRIZED PRODUCT

Let $\phi_i^{(\alpha)}$ and $\psi_j^{(\beta)}$ be two sets of functions that transform as Γ_α and Γ_β. When $\alpha = \beta$ the following manipulations show that the symmetric and antisymmetric sums also form a basis for a representation of $\Gamma_\alpha \times \Gamma_\beta$.

$$\hat{O}_s(\phi_i^{(\alpha)}\psi_j^{(\alpha)}) = \sum_{i'j'} \phi_{i'}^{(\alpha)}\psi_{j'}^{(\alpha)} D^{(\alpha)}(s)_{i'i} D^{(\alpha)}(s)_{j'j}$$

$$\hat{O}_s(\phi_j^{(\alpha)}\psi_i^{(\alpha)}) = \sum_{i'j'} \phi_{j'}^{(\alpha)}\psi_{i'}^{(\alpha)} D^{(\alpha)}(s)_{j'j} D^{(\alpha)}(s)_{i'i}$$

$$\begin{aligned} \hat{O}_s(\phi_i^{(\alpha)}\psi_j^{(\alpha)} \pm \phi_j^{(\alpha)}\psi_i^{(\alpha)}) &\qquad (13.15)\\ = \sum_{i'j'} (\phi_{i'}^{(\alpha)}\psi_{j'}^{(\alpha)} \pm \phi_{j'}^{(\alpha)}\psi_{i'}^{(\alpha)}) D^{(\alpha)}(s)_{i'i} D^{(\alpha)}(s)_{j'j} \\ = \frac{1}{2}\sum_{i',j'} (\phi_{i'}^{(\alpha)}\psi_{j'}^{(\alpha)} \pm \phi_{j'}^{(\alpha)}\psi_{i'}^{(\alpha)}) \\ \cdot (D^{(\alpha)}(s)_{i'i} D^{(\alpha)}(s)_{j'j} \pm D^{(\alpha)}(s)_{j'i} D^{(\alpha)}(s)_{i'j}) \end{aligned}$$

where the last step involves simply manipulating the next to last sum in order to achieve the more symmetrical appearing expression. Accordingly, the symmetrized product $(\phi_i^{(\alpha)}\psi_j^{(\alpha)} + \phi_j^{(\alpha)}\psi_i^{(\alpha)})$ and the antisymmetrized product $(\phi_i^{(\alpha)}\psi_j^{(\alpha)} - \phi_j^{(\alpha)}\psi_i^{(\alpha)})$ transform only into each other and form the symmetrized and antisymmetrized representations $[\Gamma_\alpha \times \Gamma_\alpha]^+$ and $[\Gamma_\alpha \times \Gamma_\alpha]^-$ for $\Gamma_\alpha \times \Gamma_\alpha$, where

$$\Gamma_\alpha \times \Gamma_\alpha = [\Gamma_\alpha \times \Gamma_\alpha]^+ + [\Gamma_\alpha \times \Gamma_\alpha]^- \qquad (13.16)$$

Since the case $i = j$ cannot occur for the antisymmetrized product (the product function difference vanishes) we find the dimensions of the two cases to be

$$\begin{aligned} d^+_{\alpha\times\alpha} &= \tfrac{1}{2}d^2_\alpha(d^2_\alpha+1) \\ d^-_{\alpha\times\alpha} &= \tfrac{1}{2}d^2_\alpha(d^2_\alpha-1) \\ d^+_{\alpha\times\alpha}+d^-_{\alpha\times\alpha} &= d^2_\alpha \end{aligned} \tag{13.17}$$

Our use of ϕ and ψ is to show the generality of the above results and also to show that when $\{\phi^{(\alpha)}_i\}=\{\psi^{(\alpha)}_j\}$; that is, when the product functions are related, the antisymmetrized products vanish identically. Thus, when the functions in a product set arise from the same identical set, only the symmetrized representation is present in the direct product of a representation itself. Thus, in the example cited earlier

$$\begin{aligned} \text{symmetrized set:}\quad & xx+xx \\ & xy+yx \qquad\qquad d^+_{E\times E}=3 \\ & yy+yy \\ \text{antisymmetrized set:}\quad & xy-yx=0; \qquad [E\times E]^- \text{ does not exist} \end{aligned}$$

The results of this section indicate that in determining whether or not an integral or matrix element vanishes one must only look at the irreducible representations contained in the *symmetrized* direct product if functions in the product set occur which arise from the same identical initial set of functions.

The irreducible representations included in the symmetrized and antisymmetrized direct products are obtained in the usual way. For the two cases, the matrices are given by [see the last equation of (13.15) and note that the unrestricted sum allows $i'\neq j'$ functions on the right-hand side to be repeated]

$$D(s)^{\pm}_{i'j',ij}=D^{(\alpha)}(s)_{i'i}D^{(\alpha)}(s)_{j'j}\pm D^{(\alpha)}(s)_{j'i}D^{(\alpha)}_{i'j} \tag{13.18}$$

In order to determine the characters of the products we set $i'=i$, $j'=j$ and use the restricted sum $i\leq j$ in order to avoid counting the product sums or differences twice

$$\begin{aligned} \chi(s)^{\pm} &= \sum_{i\leq j} D(s)^{\pm}_{ij,ij}=\tfrac{1}{2}\sum_{i,j} D(s)^{\pm}_{ij,ij} \\ &= \tfrac{1}{2}\sum_{i,j}(D^{(\alpha)}(s)_{ii}D^{(\alpha)}(s)_{jj}\pm D^{(\alpha)}(s)_{ji}D^{(\alpha)}(s)_{ij}) \\ &= \tfrac{1}{2}(\chi^{(\alpha)}(s)\chi^{(\alpha)}(s)\pm\chi^{(\alpha)}(s^2)) \end{aligned} \tag{13.19}$$

Consider again the C_{3v} example of $E\times E$.

C_{3v}	E	$2C_3$	$3\sigma_v$	
A_1	1	1	1	
A_2	1	1	−1	
E	2	−1	0	
$E\times E$	4	1	0	$=A_1+A_2+E$
$[E\times E]^+$	3	0	1	$=A_1 \quad +E$
$[E\times E]^-$	1	1	−1	$= \quad A_2$

By using the character relation given in (13.19) we see that the symmetrized direct product decomposes into A_1+E while the antisymmetrized direct product is simply A_2.

Problems

13.7 Reconsider problem 13.1 and indicate further integral vanishings when partner functions from the same set are involved ($\alpha=\beta$, $n=m$).

13.8 Indicate whether or not symmetry predicts the vanishing of the following integrals for the group $G=T_d$. The symmetry type of each function and operator is noted and repeated function symbols indicate functions from the *same* irreducible representation.

a. $(\phi^{(T_2)}, \hat{V}^{(T_1)}\psi^{(T_2)})$
b. $(\phi^{(T_2)}, \hat{V}^{(T_1)}\phi^{(T_2)})$

13.9 Consider the group C_{3v} and the functions $\phi_1=x^2-y^2$, $\phi_2=xy$ of Γ_1 and $\psi_1=xz$, $\psi_2=yz$ of Γ_2, where both Γ_1 and Γ_2 are irreducible representations for C_{3v} of type E. Show that the product set $\{\phi_i\psi_j\}$ will form a basis for a four-dimensional representation while the set $\{\psi_i\psi_j\}$ forms a basis for only a three-dimensional representation of C_{3v}.

SUMMARY

Given now the symmetry types occurring in a given integral or matrix element, we can deduce on group theoretical grounds whether the integral must vanish "by symmetry" or whether it need not be zero. The greater the detailed information we have at our disposal the more detailed will our analysis be. Knowing the row by which a symmetry adapted function transforms allows us to go further than simply knowing the irreducible representation involved. If functions from the same set of basis functions are involved we must employ the concept of symmetrized products. If no restrictions obtain the final value of the integral—zero being one of the possible outcomes—must be determined by its actual evaluation. We can use this powerful application of group theory with any functions and operators and will find it to be perhaps the single most useful application of the theory to physical systems.

SOLUTIONS TO PROBLEMS

13.1 The character table for C_{3v} is

C_{3v}	E	$2C_3$	$3\sigma_v$	
A_1	1	1	1	z
A_2	1	1	−1	
E	2	−1	0	x, y

where the entries on the right remind us that z transforms as A_1 and that x and y transform as E. Since x and y transform as the same irreducible representation we need only consider one of them. Thus, noting that here all $\Gamma_\alpha = \Gamma_\alpha^*$ we find for x

	$\phi_{n,i}^{(\alpha)}$	$\hat{V} = x$	$\phi_{m,j}^{(\beta)}$	E	$2C_3$	$3\sigma_v$	Decomposition
1.		$A_1 \times E \times A_1$		2	−1	0	E
2.		$A_1 \times E \times A_2$		2	−1	0	E
3.		$A_1 \times E \times E$		4	1	0	$A_1 + A_2 + E$
4.		$A_2 \times E \times A_2$		2	−1	0	E
5.		$A_2 \times E \times E$		4	1	0	$A_1 + A_2 + E$
6.		$E \times E \times E$		8	−1	0	$A_1 + A_2 + 3E$

while for $\hat{V} = z$ we obtain

	$\phi_{n,i}^{(\alpha)}$	$\hat{V} = z$	$\phi_{m,j}^{(\beta)}$	E	$2C_3$	$3\sigma_v$	Decomposition
1.		$A_1 \times A_1 \times A_1$		1	1	1	A_1
2.		$A_1 \times A_1 \times A_2$		1	1	−1	A_2
3.		$A_1 \times A_1 \times E$		2	−1	0	E
4.		$A_2 \times A_1 \times A_2$		1	1	1	A_1
5.		$A_2 \times A_1 \times E$		2	−1	0	E
6.		$E \times A_1 \times E$		4	1	0	$A_1 + A_2 + E$

Thus, for $\hat{V} = x$ only the integrals for cases (3), (5), and (6) *can* be nonzero while integrals for cases (1), (2), and (4) *will* be zero. For

$\hat{V}=z$, matrix elements for cases (2), (3), and (5) vanish while the others need not.

13.2 We have to show that $f_e(x)$ and $f_o(x)$ transform as different irreducible representations of some group. The group C_2 will do, although we could, if we wished, select a more complicated group. For C_2, $f_e(x)$ transforms as A while $f_o(x)$ transforms as B; thus, since $A\times B=B$ and does not contain $\Gamma_1=A$ for C_2, the matrix element (here of the unit operator) *must* vanish.

13.3 Since r_{12}^{-1} transforms as $\Gamma_1=A_{1g}$ for D_{6h} the integral will vanish unless $B_{2g}^*\times B_{1g}^*\times E_{1u}\times A_{2g}$ contains A_{1g}. The character of the multiple direct product is

	E	$2C_6$	$2C_3$	C_2	$3C_2'$	$3C_2''$	i	$2S_3$	$2S_6$	σ_h	$3\sigma_d$	$3\sigma_v$
χ	2	1	-1	-2	0	0	-2	-1	1	2	0	0

which is precisely $\chi^{(E_{1u})}$. The direct product does not contain $\Gamma_1=A_{1g}$ and thus the integral must vanish.

13.4 $\Gamma_\alpha^*\times\Gamma_\beta$ will contain Γ_1 if $c_1=(1/{}^oG)\sum_{s\in G}\chi^{(\alpha)}(s)^*\chi^{(\beta)}(s)$ is nonzero. But this expression is exactly the character orthogonality equation and vanishes unless $\Gamma_\alpha=\Gamma_\beta$. Note that $\Gamma_\alpha^*\times\Gamma_\alpha$ contains Γ_1 exactly once.

13.5 The criteria for determining whether or not such matrix elements vanish depends on whether

$$c_1=\frac{1}{{}^oG}\sum_{s\in G}\chi^{(\alpha)}(s)^*\chi^{(\eta)}(s)\chi^{(\beta)}(s)$$

is zero or not. We can write

$$\begin{aligned}\Gamma=\Gamma_\alpha^*\times\Gamma_\eta\times\Gamma_\beta&=\Gamma_\alpha^*\times(\Gamma_\eta\times\Gamma_\beta)\\&=\Gamma_\alpha^*\times\sum_\delta c_\delta\Gamma_\delta\\&=\sum_\delta c_\delta\Gamma_\alpha^*\times\Gamma_\delta\end{aligned}$$

We know from problem 13.4 that $\Gamma_\alpha^*\times\Gamma_\delta$ contains Γ_1 only if $\delta=\alpha$. Accordingly, only if $\Gamma_\eta\times\Gamma_\beta$ contains Γ_α will such matrix elements possibly be nonzero; if $\Gamma_\eta\times\Gamma_\beta$ does not contain Γ_α, the matrix element will definitely vanish.

13.6 When representations are real, $\Gamma_\alpha=\Gamma_\alpha^*$ and $\chi^{(\alpha)}(s)=\chi^{(\alpha)}(s)^*$. Thus, in a direct product like $\Gamma_\alpha\times\Gamma_\eta\times\Gamma_\beta$ the condition for Γ_1 being contained is

$$c_1\neq 0=\frac{1}{{}^oG}\sum_{s\in G}\chi^{(\alpha)}(s)\chi^{(\eta)}(s)\chi^{(\beta)}(s)$$

Γ_η is contained in $\Gamma_\alpha \times \Gamma_\beta$ if

$$c_\eta \neq 0 = \frac{1}{{}^oG} \sum_{s \in G} \chi^{(\eta)}(s)\chi^{(\alpha)}(s)\chi^{(\beta)}(s)$$

which is precisely the same expression. When real *representations are involved* we can restate the Γ_1-containing condition in a variety of ways: $\Gamma_\alpha \times \Gamma_\eta$ must contain Γ_β, $\Gamma_\alpha \times \Gamma_\beta$ must contain Γ_η, or $\Gamma_\eta \times \Gamma_\beta$ must contain Γ_α.

13.7 The same cases vanish as before for general α, β, n, and m. For $\alpha = \beta$, $n = m$ we need characterize the symmetrized and antisymmetrized products involving $\Gamma_\alpha \times \Gamma_\beta$. We have already seen in the text that $[E \times E]^+ = A_1 + E$. Thus, when $\Gamma_\alpha = \Gamma_\beta = E$, $\hat{V}$ must contain A_1 or E in order for the matrix element to be possibly nonzero. $\hat{V} = z$ transforms as A_1 while $\hat{V} = x$ (or y) transforms as E. Accordingly, the integrals in these cases are not subject to any further requirement to vanish.

For one-dimensional representations it should be clear that the direct product is one-dimensional and must be of the symmetrized type for $\Gamma_\alpha \times \Gamma_\alpha$. $A_1 \times A_1 = A_2 \times A_2 = A_1$ and there are no additional changes from our earlier analysis in problem 13.1.

Had we been considering an operator that transformed as A_2 then the $E \times E$ type integrals would be required to vanish.

13.8 a. Independent functions are involved and we can treat the usual multiple direct product. All the representations of T_d are real so we need simply find out if $T_2 \times T_2$ contains T_1. The table below shows that $T_2 \times T_2 = A_1 + E + T_1 + T_2$ so that the integral need not vanish.

b. Dependent functions are present and one must ask if $[T_2 \times T_2]^+$ contains T_1. Working out the symmetrized and antisymmetrized products we see that $[T_2 \times T_2]^+$ does not contain T_1 so that in this case the integral *must* vanish.

T_d	E	$8C_3$	$3C_2$	$6S_4$	$6\sigma_d$	
A_1	1	1	1	1	1	
A_2	1	1	1	−1	−1	
E	2	−1	2	0	0	
T_1	3	0	−1	1	−1	
T_2	3	0	−1	−1	1	
$T_2 \times T_2$	9	0	1	1	1	$A_1 + E + T_1 + T_2$
$[T_2 \times T_2]^+$	6	0	2	0	2	$A_1 + E + T_2$
$[T_2 \times T_2]^-$	3	0	−1	1	−1	T_1

13.9 Although Γ_1 and Γ_2 are both of the E type in C_{3v}, the product set $\{\phi_i\psi_j\}$ clearly consists of our independent functions:

$$\begin{Bmatrix} x^2-y^2 \\ xy \end{Bmatrix} \times \begin{Bmatrix} xz \\ yz \end{Bmatrix} = \begin{Bmatrix} (x^2-y^2)xz \\ (x^2-y^2)yz \\ x^2yz \\ xy^2z \end{Bmatrix}$$

so a basis for a four-dimensional representation is possible. In terms of the symmetrized and antisymmetrized product functions we would have

$$2\phi_1\psi_1 = 2x^3z - 2xy^2z$$

$$\phi_1\psi_2 + \phi_2\psi_1 = x^2yz - y^3z + x^2yz = 2x^2yz - y^3z$$

$$2\phi_2\psi_2 = xy^2z$$

$$\phi_1\psi_2 - \phi_2\psi_1 = x^2yz - y^3z - x^2yz = -y^3z$$

However, in the case of the product set $\{\psi_i\psi_j\}$ the functions are clearly not independent:

$$\begin{Bmatrix} xz \\ yz \end{Bmatrix} \times \begin{Bmatrix} xz \\ yz \end{Bmatrix} = \begin{Bmatrix} x^2z^2 \\ xyz^2 \\ xyz^2 \\ y^2z^2 \end{Bmatrix} \quad \text{(}xyz^2, xyz^2\text{: the same!)}$$

Or, in terms of the basis functions for $[\Gamma_2\times\Gamma_2]^+$ and $[\Gamma_2\times\Gamma_2]^-$,

$$2\psi_1\psi_1 = 2x^2z^2$$

$$\psi_1\psi_2 + \psi_2\psi_1 = xyz^2 + xyz^2 = 2xyz^2$$

$$2\psi_2\psi_2 = 2y^2z^2$$

$$\psi_1\psi_2 - \psi_2\psi_1 = xyz^2 - xyz^2 = 0$$

14
The Permutation Symmetry of Wave Functions

The past discussion of matrix elements and the wave functions involved dealt with total wave functions of a particular symmetry. That is, whatever the structure of the system involved in terms of the basic particles making it up, we know that wave functions specifying the probability amplitudes for each and every particle at any instant of time can be found and that the symmetry of the system will be reflected in the eigenfunctions of the Hamiltonian. To this point we have talked only of spatial symmetry and have not discussed another most important set of symmetry operations present in every system of identical particles—permutation symmetry. In this chapter we shall look at the permutation symmetry of quantum mechanical systems and its effect on wave functions (particularly many-electron wave functions), putting special emphasis on the orbital form of wave functions.

WAVE FUNCTIONS AND ORBITALS

A well-known result of quantum mechanics is that a system specified by a *separable* Hamiltonian can be broken up into smaller problems involving each separable part. Thus, for example, if we could write

$$\hat{\mathcal{H}} = \hat{\mathcal{H}}_a(1) + \hat{\mathcal{H}}_b(2, 3, 4) + \hat{\mathcal{H}}_c(5, 6) + \cdots \tag{14.1}$$

where $\hat{\mathcal{H}}_a(1)$ deals only with the coordinates of particle 1, $\hat{\mathcal{H}}_b(2, 3, 4)$ only with the coordinates of particle 2, 3, and 4, and so forth, then a solution to the stationary state problem $\hat{\mathcal{H}}\psi = \varepsilon\psi$ is given by

$$\begin{aligned} \psi &= \psi_a(1)\psi_b(2, 3, 4)\psi_a(5, 6) \cdots \\ \varepsilon &= \varepsilon_a + \varepsilon_b + \varepsilon_c + \cdots \end{aligned} \tag{14.2}$$

where

$$\begin{aligned} \hat{\mathcal{H}}_a(1)\psi_a(1) &= \varepsilon_a\psi_a(1) \\ \hat{\mathcal{H}}_b(2, 3, 4)\psi_b(2, 3, 4) &= \varepsilon_b\psi(2, 3, 4) \\ &\vdots \end{aligned} \tag{14.3}$$

Each of the separable problems in Eq. (14.3) may have many solutions providing us with multiple solutions for Eq. (14.2). When a system of N uncoupled identical particles is involved the identical nature of the particles requires each sub-Hamiltonian to be identical in form and we have

$$\hat{\mathcal{H}} = \sum_{i=1}^{N} \hat{\mathcal{H}}(i) \tag{14.4}$$

Each subproblem now becomes identical except for the particle labeling index so that

$$\begin{aligned} \hat{\mathcal{H}}(i)\phi_n(i) &= \varepsilon_n\phi_n(i), \qquad n = 1, 2, \ldots \\ \psi &= \phi_n(1)\phi_{n'}(2)\phi_{n''}(3)\cdots \\ \varepsilon &= \varepsilon_n + \varepsilon_{n'} + \varepsilon_{n''} + \cdots \end{aligned} \tag{14.5}$$

where now we denote the *set* of single-particle eigenfunctions by $\{\phi_n\}$, the index n being used to distinguish the various solutions. When the ϕ_n describe a spatial one-particle probability amplitude we speak of them as *orbitals* and the total wave function as an *orbital product*. The orbitals represent wave functions for the one-particle system defined by $\hat{\mathcal{H}}(i)$.

Unfortunately, life is not this simple, and the particles of interest to us do interact with each other making it generally impossible to write down separable Hamiltonians. The wave function for the two-electron helium atom would be a simple orbital product were it not for the electron-coupling coulomb interaction. The same could be said for the 92-electron uranium atom. The coulomb term is sufficiently complicating that we are currently forced into a situation where we must usually deal with it and other similar particle-coupling terms in an indirect manner. Although the problem of particle-coupling terms is a general one we shall limit our present discussion to many-electron systems in molecules of fixed nuclear geometry (i.e., the electronic part of the Bonn–Oppenheimer separation of electronic and nuclear motion). The chief villain in this situation is truly the coulomb interaction between the electrons.

First, we may simply ignore it. We may assume some effective 1-electron potential to somehow partially compensate for the effects of the electron–electron repulsion and introduce adjustable parameters into our functions to give a best fit with experiment. Despite the outcries of the quantum mechanical purists such semiempirical theories have been and are most useful in making qualitative and sometimes even semiquantitative predictions. The resulting wave functions are simple orbital products.

We may do considerably better by using the full, real Hamiltonian with the coulomb term included but employing a simple orbital product wave function. We can demand that the orbital product involved give the best (lowest) energy it can and by the usual variational techniques are led to self-consistent field methods. More work is required but we generally expect more reasonable results having used the true Hamiltonian. An orbital product wave function again is the (demanded, in this case) result.

Finally, we may recognize the fact that either through a semiempirical approach or by a self-consistent field approximation we have a set of orbitals and can form many orbital products. If our orbital set is complete, the orbital product set of wave functions constitutes a complete set for the real N-electron problem. Usually the orbital set is limited (incomplete) but hopefully adequate, giving rise to a similarly qualified set of product functions. Given such a set of orbital products we may write our total wave function as linear combinations of the various orbital products. Each orbital product is called a *configuration* and the coupling of different orbital product terms by terms such as the coulomb interaction is called *configuration interaction.* Again, orbital products are involved.

It is clear then that we can look at electronic wave functions in terms of orbital products and that the symmetry and consequences of symmetry involved will be intimately connected to the symmetry of the orbitals. Let us write then a configuration as

$$\begin{aligned}\Theta(1, 2, \ldots, N) &= \phi_a(1)\phi_b(2)\phi_c(3)\cdots\phi_n(N)\\ &\equiv a(1)b(2)c(3)\cdots\\ &\equiv abcd\cdots\end{aligned} \tag{14.6}$$

where the $\phi_a(i)$ or simply $a(i)$ represents the orbital of type a in our set and where the last defining equality indicates that a serial ordering of the particle indices is implied. Thus, for example,

$$\begin{aligned}a(1)b(3)c(2) &\equiv acb\\ a(2)b(1)c(3) &\equiv bac\end{aligned} \tag{14.7}$$

and the notation saves us from writing particle labels.

THE PERMUTATION SYMMETRY OF PHYSICAL SYSTEMS

Consider a Hamiltonian describing a system of N identical and indistinguishable particles. Since the Hamiltonian represents the energy operator for the system it should be clear that if the particles are indeed indistinguishable it cannot and must not matter how we label them. No matter what permutation of the particle indices we care to make, the Hamiltonian will not change. We see, therefore, that without any further knowledge of the system we immediately have a group of operators that commute with the Hamiltonian and constitute a group—the group of permutations on N objects, our friend S_N. According to our previous work regarding group theory and quantum mechanics we know immediately that the eigenfunctions of the Hamiltonian must—at the least—have symmetry appropriate to the irreducible representations of S_N.

For the $N=3$ problem $\Theta_1 = abc$ and $\Theta_2 = acb$, for example, are not permissible wave functions for neither has any type of permutation symmetry; nor do any of the other four orbital products *bac, bca, cab*, and *cba* which complete the set of 3! products involving the three orbitals *a, b, c*. Clearly various linear combinations of these products must be taken which possess the permutation symmetry appropriate to one of the irreducible representations of S_N. At this juncture the physical rules governing quantum mechanics simplify the situation considerably. The pertinent rule states that a unique physical state is specified once we say that identical particles occupy the orbitals *a, b*, and *c* and that wave functions like *abc* and *acb* (or any of the other four) must relate to the *same* physical situation. This means that in the present example there must exist a *unique* combination of the 3! orbital products that at the same time possesses the necessary permutation symmetry. In other words, the physical rules restrict our choice to unique, one-dimensional irreducible representations of S_N. We cannot have higher dimensionality representations for that implies a degeneracy while the physics of the situation says we are dealing with a unique physical state.

Now one of the more interesting properties of S_N is that it contains exactly two one-dimensional representations regardless of N (for $N>1$, of course). One is the always present completely symmetric (trivial) representation where $\chi=+1$ for all the permutation operators; the other is the antisymmetric representation where $\chi=+1$ for permutations that are even (made up of an even number of transpositions) and $\chi=-1$ for odd permutations (made up of an odd number of transpositions).

Problems

14.1 Show that the presence in S_N of the subgroup A_N, the alternating group of order $N!/2$, guarantees the presence of the completely symmetric and the antisymmetric irreducible representations in S_N.

14.2 Use the fact that for each and every transposition (ij) the equality $(ij)^2=1$ holds to prove that S_N has *two and only two* one-dimensional representations.

As the reader well knows, the fundamental particles of nature divide themselves precisely into two classes: bosons or Bose–Einstein particles whose multiparticle wave functions remain unchanged by any permutation and fermions or Fermi–Darac particles whose multiparticle wave functions change sign upon odd permutation but are unaffected by even permutations. Thus, wave functions for identical boson systems must transform as the completely symmetric representation of S_N while wave functions of identical fermion systems must transform as the antisymmetric representation of S_N. These results are facts of nature and, besides being consistent with our general understanding of the relationship of group theory and quantum mechanics (which, of course, it must be!), afford no further simplification of our physical problem. Such being the case, we accept the required symmetry or antisymmetry of the wave functions we must work with and do not consider further the permutation symmetry of the system in terms of the total symmetry group of the Hamiltonian. The consequences of the presence of the permutation symmetry does indeed emerge, however, in our further group theoretical considerations.

Since a multiparticle wave function must transform as Γ_S (completely symmetric representation) or as Γ_A (antisymmetric representation), we may use the standard group projection operators to construct wave functions of the proper symmetry. Since the projection operator for the αth irreducible representation is

$$\hat{P}^{(\alpha)} = \frac{d_\alpha}{{}^oG} \sum_{s \in G} \chi^{(\alpha)}(s)^* \hat{O}_s \tag{14.8}$$

we have for Γ_S and Γ_A

$$\begin{aligned} \hat{P}^{(\Gamma_S)} &\equiv \hat{\mathcal{S}} = \frac{1}{N!} \sum_i \hat{p}_i \qquad (\Gamma_S) \\ \hat{P}^{(\Gamma_A)} &\equiv \mathcal{A} = \frac{1}{N!} \sum_i s_i \hat{p}_i \qquad (\Gamma_A) \end{aligned} \tag{14.9}$$

where $\hat{p}_i$ represents the ith permutation in S_N. s_i is the signature of $\hat{p}_i$, being $+1$ for even permutations and -1 for odd permutations, and $\hat{\mathcal{S}}$ and $\hat{\mathcal{A}}$ are the usual symbols employed for $\hat{P}^{(\Gamma_S)}$ and $\hat{P}^{(\Gamma_A)}$. Thus, since the configuration $\Theta = abc$ represents a unique physical situation the (unnormalized) acceptable wave functions describing the physical situation would be

$$\begin{aligned} \psi_S &= \hat{\mathcal{S}}\Theta = \frac{1}{N!} \sum_i \hat{p}_i \Theta \qquad \text{(bosons)} \\ \psi_A &= \hat{\mathcal{A}}\Theta = \frac{1}{N!} \sum_i s_i \hat{p}_i \Theta \qquad \text{(fermions)} \end{aligned} \tag{14.10}$$

For a three-identical-boson problem we obtain (taking the sum of permutation operators for S_3 in the order e, (12), (13), (23), (123), (132))

$$\psi_S = \tfrac{1}{6}(abc + bac + cba + acb + cab + bca) \tag{14.11}$$

while for a three-identical-fermion problem we would get

$$\psi_A = \tfrac{1}{6}(abc - bac - cba - acb + cab + bca) \tag{14.12}$$

As the reader may well know and can easily verify, a function such as ψ_A may be written in determinantal form (the Slater determinant):

$$\psi_A = \tfrac{1}{6}\begin{vmatrix} a(1) & b(1) & c(1) \\ a(2) & b(2) & c(2) \\ a(3) & b(3) & c(3) \end{vmatrix} \tag{14.13}$$

The generalization to any N is straightforward and yields

$$\psi_A(1, 2, \ldots, N) = \frac{1}{N!}\begin{vmatrix} a(1) & b(1) & \cdots & n(1) \\ a(2) & b(2) & \cdots & n(2) \\ \cdot & \cdot & & \cdot \\ \cdot & \cdot & & \cdot \\ \cdot & \cdot & & \cdot \\ a(N) & b(N) & \cdots & n(N) \end{vmatrix} \tag{14.14}$$

We may always normalize our wave functions and shall mention this matter briefly later.

FERMIONS AND THE EXCLUSION PRINCIPLE

We now restrict ourselves to systems of electrons (fermions) with fixed nuclei. In order to specify completely an electron's situation in an orbital, we must specify its spin; the combination of the spatial plus spin characteristics is sometimes called a *spin orbital*. The spin of each electron can be "up" (α, $m_s = +\frac{1}{2}$) or "down" (β, $m_s = -\frac{1}{2}$) and the corresponding spin orbitals are conveniently labeled as a (α spin) or $\bar{a}$ (β spin). Thus, a configuration with spin now specified might be $\Theta = a\bar{a}b$ corresponding to electron #1 in an a-type spatial orbital with spin α, #2 in an a-type spatial orbital with spin β, and #3 in a b-type orbital with spin α. Let us denote the various spin orbitals as μ_i so that, for example, if $\mu_1 = a$, $\mu_2 = \bar{a}$, $\mu_3 = b, \ldots$ the above configuration $\Theta = a\bar{a}b$ becomes $\Theta = \mu_1\mu_2\mu_3$. In general, then, given a spatial basis $\{a, b, c, \ldots\}$ we can easily form a spin orbital basis $\{\mu_1, \mu_2, \ldots\}$ by combining each orbital with α and β spin characteristics. A general N-electron configuration can then be written as $\Theta = \mu_i\mu_j\mu_k \cdots$ (up to N spin orbitals) and a properly antisymmetrized (but not necessarily normalized) wave function as

$$\psi = \hat{\mathscr{A}}\Theta \tag{14.15}$$

The nature of the $\hat{\mathscr{A}}$ projection operator is such that any pairwise permutation (transposition) of a properly antisymmetrized electronic wave function causes the function to change sign.

Problems

14.3 Show that (ij) and $\hat{\mathscr{A}}$ commute

$$(ij)\hat{\mathscr{A}} = \hat{\mathscr{A}}(ij)$$

14.4 Show that $\hat{\mathscr{A}}$ and any transpositions (ij) obey the relation

$$(jl)\hat{\mathscr{A}} = -\hat{\mathscr{A}}$$

so that

$$(jl)\psi = (jl)\hat{\mathscr{A}}\Theta = -\hat{\mathscr{A}}\Theta = -\psi$$

The above-mentioned property gives rise to the famous Pauli exclusion principle: No two fermions may occupy the same spin orbital. That is, two fermions may have the same orbital description but they must then have opposite (different) spin. That this is so should be clear from the application of (ij) to a antisymmetrized spin orbital product containing a doubly occupied spin orbital. Thus, let

$$\psi = \hat{\mathscr{A}}\mu\mu\nu \tag{14.16}$$

Employing the results of problem 14.3 as applied to this particular wave function, we get

$$\begin{aligned}(12)\psi = (12)\hat{\mathscr{A}}\mu\mu\nu &= \hat{\mathscr{A}}(12)\mu\mu\nu \\ &= \hat{\mathscr{A}}\mu\mu\nu \\ &= \psi \end{aligned} \tag{14.16}$$

yet from problem 14.4 we have

$$\begin{aligned}(12)\psi = (12)\hat{\mathscr{A}}\mu\mu\nu &= -\hat{\mathscr{A}}\mu\mu\nu \\ &= -\psi \end{aligned} \tag{14.17}$$

The only way both (14.16) and (14.17) can hold is if $\psi = -\psi = 0$. That is, $\hat{\mathscr{A}}$ acting on $\mu\mu\nu$ annihilates the function and the result is no wave function at all.

The result is easily generalized to any number of fermions. Wave functions for identical fermions must change sign upon the interchange of any two particle labels. For boson systems similar reasoning leads to the result that such an interchange must leave the multiple boson wave function unaffected.

MATRIX ELEMENTS AND ANTISYMMETRIZED ORBITAL PRODUCTS

Permutation operators are unitary operators in that $(\hat{p}_i f, \hat{p}_i g) = (f, g)$ for any two functions f, g. The projection operators $\hat{P}^{(\alpha)}$ for a unitary group are unitary; indeed, we can show that they are also hermitian, a characteristic that, as we shall see, is extremely useful in evaluating matrix elements.

Problems

14.5 Show that the adjoint of a projection operator $\hat{P}$ (having the same domain as $\hat{P}$) is also a projection operator.

14.6 Show that the group projection operators $\hat{P}^{(\alpha)}$ (see Chapter 11) for finite unitary groups are hermitian.

The permutation group is clearly a unitary group; problem 14.6 shows us then that $\hat{\mathscr{A}}$, being a group projection operator of a unitary group, is hermitian. The hermitian character of the antisymmetrization operator is conveniently used to correctly normalize an electronic orbital product. Let

$$\psi = k\hat{\mathscr{A}}\Theta = k\hat{\mathscr{A}}\mu_1\mu_2\cdots\mu_N \tag{14.18}$$

be a properly antisymmetric wave function for an N-electron system. k is the factor necessary to normalize ψ, and we shall assume a set of orthonormal spin orbitals: $(\mu_i, \mu_j) = \delta_{ij}$. Then

$$\begin{aligned} 1 = (\psi, \psi) &= k^2(\hat{\mathscr{A}}\Theta, \hat{\mathscr{A}}\Theta) \\ &= k^2(\Theta, \hat{\mathscr{A}}^2\Theta) \\ &= k^2(\Theta, \hat{\mathscr{A}}\Theta) \end{aligned} \tag{14.19}$$

using both the hermitian and projection operator characteristics of $\hat{\mathscr{A}}$ in that order. The terms in the integral

$$(\Theta, \hat{\mathscr{A}}\Theta) = \left(\mu_1\mu_2\cdots\mu_N, \frac{1}{N!}\sum_i s_i\hat{p}_i\mu_1\mu_2\cdots\mu_N\right) \tag{14.20}$$

involving any permutation other than e will vanish due to the orthonormality of the μ_i, so that we get finally

$$\begin{aligned} 1 &= k^2\left(\Theta, \frac{1}{N!}\Theta\right) \Rightarrow k = \sqrt{N!} \\ \psi &= \sqrt{N!}\hat{\mathscr{A}}\Theta \end{aligned} \tag{14.21}$$

The product $\sqrt{N!}\hat{\mathscr{A}}$ is sometimes referred to as the antisymmetrization and renormalization operator.

The same type of manipulation is involved in the simplification of general matrix elements. Since the operators referring to the identical particles of the system are symmetric in all the particle labels they commute with all permutations and, in particular, with $\hat{\mathscr{A}}$. Let $\hat{V}$ be such an operator whose matrix elements are desired between orthonormal spin orbital products:

$$\begin{aligned} V_{12} = (\psi_1, \hat{V}\psi_2) &= (\sqrt{N!}\,\hat{\mathscr{A}}\Theta_1, \hat{V}\sqrt{N!}\,\hat{\mathscr{A}}\Theta_2) \\ &= N!\,(\hat{\mathscr{A}}\Theta_1, \hat{V}\hat{\mathscr{A}}\Theta_2) \\ &= N!\,(\Theta_1, \hat{V}\hat{\mathscr{A}}\Theta_2) \\ &= \left(\Theta_1, \hat{V}\sum_i s_i\hat{p}_i\Theta_2\right) \end{aligned} \tag{14.22}$$

When we considered the normalization problem $\hat{V}$ was simply the unit operator for each particle. If $\hat{V} = 1$ here the matrix element will vanish unless Θ_1 and Θ_2 contain the same spin orbitals. If $\hat{V}$ consists of a sum $\sum_j \hat{h}^{(1)}(j)$ of 1-electron operators V_{12} will vanish if Θ_1 and Θ_2 differ by more than one spin orbital, while if $\hat{V}$ is a sum of 2-electron terms $\sum_{i<j} \hat{h}^{(2)}(i, j)$, Θ_1 and Θ_2 may differ by no more than two spin orbitals in order to be possibly nonvanishing. In the first case of 1-electron operators and two configurations differing by one spin orbital (μ_i and μ_j) V_{12} reduces to $s_k(\mu_i, \hat{h}^{(1)}\mu_j)$ where s_k is the sign of the permutation necessary to place the other spin orbitals in Θ_2 in proper order to integrate out with their partners in Θ_1. $s_k = \pm 1$ does not affect the selection rule and we see that the symmetry rules in this case may be applied to the 1-electron integral $(\mu_i, \hat{h}^{(1)}\mu_j)$ rather than the full V_{12} integral. The selection rules derived earlier involving the full wave functions ψ_1 and ψ_2 and the operator $\hat{V}$ are still valid, of course, but it will usually be simpler to treat the less complicated one-spin-orbital integral.

Part of the advantage of the above reduction is concerned with the care that must be taken in assigning a symmetry to a many-particle wave function based on the symmetry of the orbitals involved. Thus, given the symmetry adapted spin orbitals $\mu_i^{(\alpha)}$ transforming as Γ_α in G and the orbital product

$$\Theta = \mu_1^{(\alpha)}\mu_2^{(\beta)}\cdots\mu_N^{(\gamma)}$$

it is true that Θ transforms as

$$\Gamma_\alpha \times \Gamma_\beta \times \cdots \times \Gamma_\gamma$$

but it is *not* necessarily true that $\hat{\mathscr{A}}\Theta$ transforms the same way. The reason is that when $\hat{O}_a \in G$ operates in Θ some new Θ's *may* be generated that violate the Pauli principle. Thus, while $\Gamma_\alpha \times \Gamma_\beta \times \cdots \times \Gamma_\gamma$ will be a representation of dimension $d_\alpha d_\beta \cdots d_\gamma$, the $\hat{\mathscr{A}}$ projection operator may well annihilate some of the terms resulting from $\hat{O}_a\Theta$ so that a lower-dimensionality representation may result. For example, consider $G = C_{3v}$ and

$$\begin{aligned} \psi_1 &= \sqrt{N!}\,\hat{\mathscr{A}}a_1^{(E)}a_2^{(E)} \\ \psi_2 &= \sqrt{N!}\,\hat{\mathscr{A}}a_1^{(E)}\bar{a}_2^{(E)} \end{aligned} \tag{14.23}$$

where a_1 and a_2 are partner functions of an E-type representation and where we revert to the space-plus-spin notation to illustrate an effect of the latter. ψ_1 does not transform as $E\times E = A_1 + A_2 + E$ because we know that placing two electrons with parallel (the same) spin in the partners of a twofold degenerate orbital set is a unique situation; ψ_1 must, therefore, transform as some one-dimensional representation of C_{3v}. Were a_1 and a_2 functions from *different* E-representations, ψ_1 would indeed transform as $E\times E$ since in this case there are indeed four different and nonvanishing antisymmetric wave functions possible based on $a_1^{(E_1)}a_1^{(E_2)}$, $a_1^{(E_1)}a_2^{(E_2)}$, $a_2^{(E_1)}a_1^{(E_2)}$, and $a_2^{(E_1)}a_2^{(E_2)}$. For ψ_2 there again is no special problem; the spins differ and even when partner functions of the same E representation are involved four different and allowable orbital products can be used: $a_1^{(E)}\bar{a}_1^{(E)}$, $a_1^{(E)}\bar{a}_2^{(E)}$, $a_2^{(E)}\bar{a}_1^{(E)}$, and $a_2^{(E)}\bar{a}_2^{(E)}$.

Note finally that, in worrying about the assignment of symmetry to a wave function, we need not concern ourselves here with the problem of dependent products treated in the previous chapter. The reason is that each orbital function carries with it the label (*different* for each) of one of the particles in the system and each is, therefore, linearly independent of the others.

Problem

14.7 Consider $G = C_{3v}$ and the four spatial orbitals $a^{(A_1)}$, $a^{(A_2)}$, $a_1^{(E)}$, and $a_2^{(E)}$, the latter two orbitals being partner functions in the same E representation. Classify the symmetry of the following (unnormalized) wave functions.

a. $\hat{\mathscr{A}}a^{(A_1)}a^{(A_2)}$
b. $\hat{\mathscr{A}}a^{(A_1)}\bar{a}^{(A_2)}$
c. $\hat{\mathscr{A}}a^{(A_2)}\bar{a}^{(A_2)}$
d. $\hat{\mathscr{A}}a^{(A_2)}\bar{a}_1^{(E)}$
e. $\hat{\mathscr{A}}a^{(A_2)}a_1^{(E)}\bar{a}_2^{(E)}$
g. $\hat{\mathscr{A}}a_1^{(E)}a_2^{(E)}$
h. $\hat{\mathscr{A}}a_1^{(E)}\bar{a}_1^{(E)}a_2^{(E)}$
i. $\hat{\mathscr{A}}a_1^{(E)}\bar{a}_1^{(E)}a_2^{(E)}\bar{a}_2^{(E)}$

Problem 14.7 shows that physical arguments often save us much work when determining the overall symmetry of an electronic wave function. In general, a shell of orbitals of symmetry type α and degeneracy d (dimension of Γ_α) filled with n electrons—the configuration α^n—will have the same symmetry as the configuration α^{d-n}; that is, the "hole" configuration transforms as the equivalent electron configuration (see problem 14.7h). Furthermore, any closed-shell configuration α^d will always transform as the completely symmetric representation of the group (see problem 14.7i). Accordingly, when determining the symmetry of a configuration all closed shells contribute effectively A_1 or A_{1g} or whatever the completely symmetric representation notation is to the direct product under consideration.

Problems

14.8 Use the above rules to determine the symmetry of the following configurations. Orbital states referring to different sets of functions belonging to the same symmetry type are distinguished by primes. Lowercase letters are used in the designation of the orbital symmetry type.

a. $G = D_2$ $(a)^2(b_1)^2$
b. $G = D_2$ $(a^2)(b_1)(b_2)$
c. $G = C_{4v}$ $(a_1)^2(a_2)^2(b_1)(e)$
d. $G = C_{4v}$ $(a_1^2)(e)$
e. $G = C_{4v}$ $(a_1^2)(e)^3$
f. $G = D_{6h}$ $(a_{1g})^2(e_{1g})^4$
g. $G = D_{3d}$ $(a_{1g})^2(e_g)^3(e_{\mu})$
h. $G = D_{3d}$ $(a_{1g})(a_{1u})(e_g)(e_u)$
i. $G = D_{6h}$ $(a_{1g})^2(b_{2g})(b_{2u})^2$

14.9 The 1-electron π molecular orbitals of naphthalene ($G = D_{2h}$) have the following order of increasing energy:

$$b_{1u} < b_{3g} < b_{2g} < b'_{1u} < a_u < b'_{3g} < b'_{2g} < b''_{1u} < a'_u < b''_{2g}$$

so that the ground-state configuration is $(b_{1u})^2(b_{2g})^2(b_{3g})^2(b'_{1u})^2(a_u)^2$. Consider only singly excited configurations arising from exciting an a_u electron and deduce the electric dipole selection rules for the π–π^* transitions in naphthalene. (See Figure 1.)

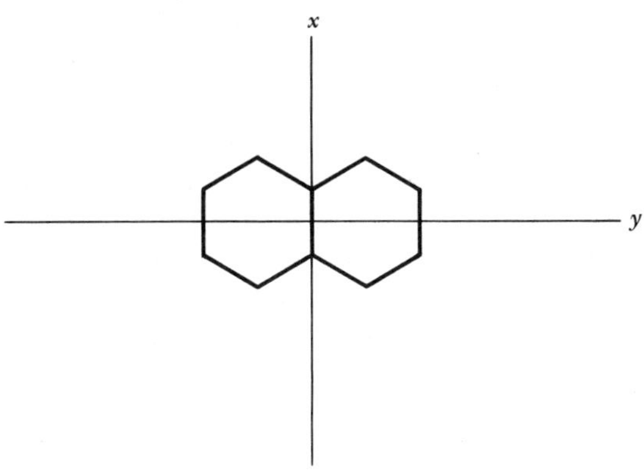

Figure 1

14.10 Suppose it was desired to carry out a configuration interaction calculation for the π-system of naphthalene described in problem 14.9 using only singly excited configurations. What singly excited configurations would mix with the ground-state configuration? What other singly excited configurations would mix with the first excited configuration $(b_{1u})^2(b_{2g})^2(b_{3g})^2(b'_{1u})^2(a_u)(b'_{3g})$?

SOLUTIONS TO PROBLEMS

14.1 The alternating group $A_N \subset S_N$ contains all $N!/2$ even permutations in S_N and is a subgroup of index 2. Being of index 2, A_N is a normal subgroup and hence serves as the kernel of the mapping between S_N and the factor group S_N/A_N isomorphic to C_2 (or S_2). Since a representation results from a mapping and since any homomorphic image of a group G is isomorphic to some factor group G/N (N being a normal subgroup), the presence of the factor group $S_N/A_N = C_2$ guarantees the representations resulting from the mappings to be

1. $A_N \rightarrow +1$

   $S_N - A_N \rightarrow +1$

2. $A_N \rightarrow +1$

   $S_N - A_N \rightarrow -1$

14.2 Since $(ij)^2 = e$, it follows that those permutations in S_N that are simple (single transpositions) must in any one-dimensional representations have characters of ± 1 since $\chi(e) = +1$ always. The characters in a one-dimensional representation are the matrix representations and obey the group algebra; thus, since every permutation can be written as the product of transpositions, it follows that every character can only be ± 1. All permutations having the same cycle structure—in particular, all transpositions—belong to the same class and must have the same character. Accordingly, if the character of (ij) is $+1$ *all* permutations must have $\chi = +1$ and we have the completely symmetric representation. If the character of (ij) is -1 then all even permutations will have $\chi = +1$ (being an even multiple of -1) while odd permutations will have $\chi = -1$ (being an odd multiple of -1), and we have the antisymmetric representation. These two representations result from the only two choices for the character of (ij) and constitute the two and only two one-dimensional representations of S_N.

14.3 The operator $\hat{\mathscr{A}}$ given by

$$\hat{\mathscr{A}} = \frac{1}{N!}\sum_i s_i\hat{p}_i = \frac{1}{N!}\Big(\sum_{\text{even}} \hat{p}_i - \sum_{\text{odd}} \hat{p}_i\Big)$$

is the positive sum of all even permutations added to the negative sum of all the odd permutations, the total being divided by $N!$ Since any $\hat{p}_k$ multiplied by a single transposition (ij) changes its parity and since there are an equal number of each parity type (recall the alternating group) it is clear that

$$(ij)\cdot\Big(\sum_{\text{even}} \hat{p}_k\Big) = \Big(\sum_{\text{odd}} \hat{p}_k\Big)$$

$$\Big(\sum_{\text{even}} \hat{p}_k\Big)\cdot(ij) = \Big(\sum_{\text{odd}} \hat{p}_k\Big)$$

with similar expressions for the product of (ij) and $(\sum_{\text{odd}} p_k)$. The commutation of (ij) and either the odd or even sum is clear so that

$$(ij)\hat{\mathscr{A}} = \hat{\mathscr{A}}(ij)$$

14.4 Using the initial result of problem 14.3 we have

$$(ij)\Big\{\sum_{\text{even}} p_k - \sum_{\text{odd}} p_k\Big\} = \Big\{\sum_{\text{odd}} p_k - \sum_{\text{even}} p_k\Big\}$$

$$= -\Big\{\sum_{\text{even}} p_k - \sum_{\text{odd}} p_k\Big\}$$

so that

$$(ij)\hat{\mathscr{A}} = \hat{\mathscr{A}}(ij) = -\hat{\mathscr{A}}$$

14.5 A projection operator satisfies the equation $\hat{P}^2 = \hat{P}$. Thus, since for arbitrary u, v

$$(u, \hat{P}^2 v) = ((\hat{P}^\dagger)^2 u, v)$$

$$(u, \hat{P}^2 v) = (u, \hat{P}v) = ((\hat{P}^\dagger)u, v)$$

and $\hat{P}^\dagger = (\hat{P}^\dagger)^2$ so that $\hat{P}^\dagger$, the adjoint of $\hat{P}$, is also a projection operator. We are allowed this type of manipulation provided P and $P^\dagger$ have the same domain; that is, their operations are definable on the same set of functions in the Hilbert space in question and result in functions contained in the Hilbert space.

14.6 The group projection operators based on the characters are given by

$$\hat{P}^{(\alpha)} = \frac{d_\alpha}{{}^oG}\sum_{s\in G}\chi^{(\alpha)}(s)^*\hat{O}_s$$

Using that fact that $\chi(s^{-1}) = \chi(s)^*$ for finite groups and given that the operator group is unitary ($\hat{O}_s^\dagger = \hat{O}_s^{-1}$) it follows that

$$\begin{aligned}\hat{P}^{(\alpha)\dagger} &= \frac{d_\alpha}{{}^oG} \sum_{s \in G} \chi^{(\alpha)}(s) \hat{O}_s^\dagger \\ &= \frac{d_\alpha}{{}^oG} \sum_{s \in G} \chi^{(\alpha)}(s) \hat{O}_s^{-1} \\ &= \frac{d_\alpha}{{}^oG} \sum_{s^{-1} \in G} \chi^{(\alpha)}(s^{-1})^* \hat{O}_{s^{-1}} \\ &= \hat{P}^{(\alpha)}\end{aligned}$$

14.7 a. No restrictions, $\Gamma = A_1 \times A_2 = A_2$.

b. Spin has no effect here so again $\Gamma = A_1 \times A_2 = A_2$.

c. $\Gamma = A_2 \times A_2 = A_1$.

d. $\Gamma = A_2 \times E = E$. The 2-electron partner function of the E representation is $\hat{\mathscr{A}} a^{(A_2)} \bar{a}_2^{(E)}$.

e. Since electrons "2" and "3" have opposite spin the correct symmetry is the mixture $\Gamma = A_2 \times E \times E = A_1 + A_2 + E$.

g. This is a unique orbital product since the spins are the same. Since it is unique it must transform as a one-dimensional representation. A_1 and A_2 differ in their characters under the operations σ_v. Thus, referring to the E representation given in problem 8.4

$\mathbf{D}^{(E)}(\sigma_1)$	$\mathrm{D}^{(E)}(\sigma_2)$	$\mathrm{D}^{(E)}(\sigma_3)$
$\begin{pmatrix} -1 & 0 \\ 0 & 1 \end{pmatrix}$	$\begin{pmatrix} 1/2 & -\sqrt{3}/2 \\ -\sqrt{3}/2 & -1/2 \end{pmatrix}$	$\begin{pmatrix} 1/2 & \sqrt{3}/2 \\ \sqrt{3}/2 & -1/2 \end{pmatrix}$

and considering σ_1 we obtain

$$\hat{\sigma}_1(a_1^{(E)}, a_2^{(E)}) = (a_1^{(E)}, a_2^{(E)}) \begin{pmatrix} -1 & 0 \\ 0 & 1 \end{pmatrix} = (-a_1^{(E)}, a_2^{(E)})$$

so that

$$\begin{aligned}\hat{\sigma}_1 \hat{\mathscr{A}} a_1^{(E)} a_2^{(E)} &= \hat{\mathscr{A}} \hat{\sigma}_1 a_1^{(E)} a_2^{(E)} \\ &= -\hat{\mathscr{A}} a_1^{(E)} a_2^{(E)}\end{aligned}$$

and we see that the wave function transforms as A_2. It cannot make any difference which σ_v we employ since they all belong to the same class. If we try, for example, σ_2 we find

$$\begin{aligned}\hat{\sigma}_2(a_1^{(E)}, a_2^{(E)}) &= (a_1^{(E)}, a_2^{(E)}) \begin{pmatrix} 1/2 & -\sqrt{3}/2 \\ -\sqrt{3}/2 & -1/2 \end{pmatrix} \\ &= \left(\tfrac{1}{2} a_1^{(E)} - \frac{\sqrt{3}}{2} a_2^{(E)}, -\frac{\sqrt{3}}{2} a_1^{(E)} - \tfrac{1}{2} a_2^{(E)} \right)\end{aligned}$$

so that

$$\hat{\mathscr{A}}\hat{\sigma}_2 a_1^{(E)} a_2^{(E)} = \hat{\mathscr{A}}\left\{\left(\tfrac{1}{2}a_1^{(E)} - \frac{\sqrt{3}}{2}a_2^{(E)}\right)\left(-\frac{\sqrt{3}}{2}a_1^{(E)} - \tfrac{1}{2}a_2^{(E)}\right)\right\}$$
$$= \hat{\mathscr{A}}\{-\tfrac{1}{4}a_1^{(E)}a_2^{(E)} + \tfrac{3}{4}a_2^{(E)}a_1^{(E)}\}$$
$$= -\hat{\mathscr{A}}a_1^{(E)}a_2^{(E)}$$

having used the fact that $\hat{\mathscr{A}}$ annihilates terms such as $a_1^{(E)}a_1^{(E)}$ and that $\hat{\mathscr{A}}a_2^{(E)}a_1^{(E)} = -\hat{\mathscr{A}}a_1^{(E)}a_2^{(E)}$.

h. Quantum mechanics tells us that there are two physically distinct but energetically degenerate ways of placing three electrons in a spatially doubly degenerate orbital so that we would expect this wave function to transform as E on physical grounds. We might argue that the symmetry is $A_1 + A_2$ which would also have two states associated with it, although group theory would not predict an energy degeneracy in such a case.

To confirm our choice of E we can proceed in several ways. We could use the C_{3v} group projection operators acting on the wave function to see indeed what symmetries are projected out (or annihilated), or since the choice is apparently between $A_1 + A_2$ and E, we could find the partner function and check the character involved with the operation C_3 (or $\bar{C}_3$), since the characters differ in the two cases for this operator. The latter method is simpler in this instance.

The partner function to $\hat{\mathscr{A}}a_1^{(E)}\bar{a}_1^{(E)}a_2^{(E)}$ must be $\hat{\mathscr{A}}a_1^{(E)}a_2^{(E)}\bar{a}_2^{(E)}$, for that is the only choice (to within a sign). Under C_3 we have

$$\hat{C}_3(\hat{\mathscr{A}}a_1^{(E)}\bar{a}_1^{(E)}a_2^{(E)}) = \hat{\mathscr{A}}\left\{\left(-\tfrac{1}{2}a_1^{(E)} - \frac{\sqrt{3}}{2}a_2^{(E)}\right)\right.$$
$$\times\left.\left(-\tfrac{1}{2}\bar{a}_1^{(E)} - \frac{\sqrt{3}}{2}\bar{a}_2^{(E)}\right)\left(\frac{\sqrt{3}}{2}a_1^{(E)} - \tfrac{1}{2}a_2^{(E)}\right)\right\}$$
$$= \hat{\mathscr{A}}a_1^{(E)}\bar{a}_1^{(E)}a_2^{(E)}[-\tfrac{1}{8} - \tfrac{3}{8}]$$
$$+ \hat{\mathscr{A}}a_1^{(E)}a_2^{(E)}\bar{a}_2^{(E)}\left[+\frac{\sqrt{3}}{8} + \frac{3\sqrt{3}}{8}\right]$$

and

$$\hat{C}_3(\hat{\mathscr{A}}a_1^{(E)}a_2^{(E)}\bar{a}_2^{(E)}) = \hat{\mathscr{A}}\left(-\tfrac{1}{2}a_1^{(E)} - \frac{\sqrt{3}}{2}a_2^{(E)}\right)\left(\frac{\sqrt{3}}{2}a_1^{(E)} - \tfrac{1}{2}a_2^{(E)}\right)$$
$$\times\left(\frac{\sqrt{3}}{2}\bar{a}_1^{(E)} - \tfrac{1}{2}\bar{a}_2^{(E)}\right)$$
$$= \hat{\mathscr{A}}a_1^{(E)}\bar{a}_1^{(E)}a_2^{(E)}\left[-\frac{\sqrt{3}}{8} - \frac{3\sqrt{3}}{8}\right]$$
$$+ \hat{\mathscr{A}}a_1^{(E)}a_2^{(E)}\bar{a}_2^{(E)}[-\tfrac{1}{8} - \tfrac{3}{8}]$$

The character involved is $-\frac{1}{2}-\frac{1}{2}=-1$ indicating that the symmetry is indeed E.

i. This is a unique physical state and must transform as a one-dimensional representation. Applying the σ_2 test as in part g we find

$$\begin{aligned}\sigma_2\hat{\mathscr{A}}a_1^{(E)}\bar{a}_1^{(E)}a_2^{(E)}\bar{a}_2^{(E)} &= \hat{\mathscr{A}}\left(\tfrac{1}{2}a_1^{(E)}-\frac{\sqrt{3}}{2}a_2^{(E)}\right)\cdot\left(\tfrac{1}{2}\bar{a}_1^{(E)}-\frac{\sqrt{3}}{2}\bar{a}_2^{(E)}\right)\\ &\quad\cdot\left(-\frac{\sqrt{3}}{2}a_1^{(E)}-\tfrac{1}{2}a_2^{(E)}\right)\cdot\left(-\frac{\sqrt{3}}{2}\bar{a}_1^{(E)}-\tfrac{1}{2}\bar{a}_2^{(E)}\right)\\ &= \hat{\mathscr{A}}a_1^{(E)}\bar{a}_1^{(E)}a_2^{(E)}\bar{a}_2^{(E)}[+\tfrac{1}{16}+\tfrac{3}{16}+\tfrac{3}{16}+\tfrac{9}{16}]\\ &= +1\hat{\mathscr{A}}a_1^{(E)}\bar{a}_1^{(E)}a_2^{(E)}\bar{a}_2^{(E)}\end{aligned}$$

so that the wave function transforms as A_1, the completely symmetric representation.

14.8 a. Closed shells; $\Gamma = A$.

b. $\Gamma = B_1\times B_2 = B_3$. This would be an excited state of the configuration in (a).

c. $\Gamma = B_1\times E = E$.

d. $\Gamma = E$.

e. $\Gamma = E$; the "hole" configuration is $(e)^1$.

f. Closed shells; $\Gamma = A_{1g}$.

g. $\Gamma = E_g\times E_\mu = A_{1u}+A_{2u}+E_u$.

h. $\Gamma = A_{1u}\times E_g\times E_u = A_{1u}\times(A_{1u}+A_{2u}+E_u)$

$$= A_{1g}+A_{2g}+E_g$$

i. $\Gamma = B_{2g}$.

14.9 The possible excitations are

$$a_\mu \rightarrow b'_{3g},\ b'_{2g},\ b''_{1u},\ a'_u,\ b''_{2g}$$

Since the matrix elements involved refer to the 1-electron operators x, y, or z we need only consider integrals such as $(a_u\times b'_{3g})$, for example. Since $\Gamma_x = B_{3u}$, $\Gamma_y = B_{2u}$, $\Gamma_z = B_{1u}$ and since our representations are one-dimensional, a transition $\pi_1\rightarrow\pi_2$ involving q ($q = x, y, z$) will be allowed only if $\Gamma_{\pi 1}\times\Gamma_{\pi 2} = \Gamma_q$. Thus,

$A_\mu\times B_{3g} = B_{3u}$	x-polarized
$A_\mu\times B_{2g} = B_{2u}$	y-polarized
$A_\mu\times B_{1u} = B_{1g}$	forbidden
$A_u\times A_u = A_g$	forbidden

Note that since Γ_q is u (all x, y, z) we could immediately see that $\mu\rightarrow\mu$ transitions ($a_\mu\rightarrow b''_{1u}$, $a_\mu\rightarrow a'_\mu$) will not be allowed.

14.10 The ground-state configuration has the symmetry A_g since it consists of closed shells. Configuration involving the excitation $a_\mu \to \pi_i$ must then be such that $A_\mu \times \Gamma_{\pi_i} = A_g$. Only $A_\mu \times A_\mu = A_g$ so only the excited configuration $(b_{1u})^2(b_{2g})^2(b_{3g})^2(b'_{1u})^2(a_u)(a'_u)$ can by symmetry mix with the ground state. Remember the matrix element under consideration involves the Hamiltonian itself which, of course, transforms as A_g.

The first excited configuration has the symmetry $A_u \times B_{3g} = B_{3u}$. There are *no* other singly excited configurations of the type $a_\mu \to \pi_i$ of this symmetry so that one might expect the first excited configuration to be a good approximation to the true first π excited state.

15
Orbital Approaches to Chemical Bonding

In the usual orbital approach to chemical problems there are basically two types of approximations—the molecular orbital (MO) and valence bond (VB) approximations. In each case we run the gamut of sophistication; in the MO approach we can start with the simple Hückel 1-electron approach and work up to the Roothaan LCAO-MO self-consistent field, while VB treatments may be of the nearest-neighbor-equal-exchange type or the more rigorous extended valence bond method. In either major category of approximate theory, group theory and the symmetry of wave functions play important roles. In fact, in many cases it is the presence of symmetry and the restrictions it places on physical observables that allows the approximate treatment to work (be useful) as well as it does. We will examine aspects of both of these methods with an emphasis on the group theory part rather than the specific method itself. Indeed, in order not to miss the forest because of the trees it is often advantageous to illustrate the use of group theory with the simpler quantum mechanical models.

MOLECULAR ORBITAL THEORY

Simple molecular orbital theory is a 1-electron approach where the pertinent integrals that arise are suitably parameterized so as to give wave functions and energies that mimic those obtained by more rigorous treatments. Because of its simplistic nature it is often looked down upon by those who employ more sophisticated techniques. The fact is, however, it has been *useful* and is a *useable* theory, and when properly leashed, is capable of pointing us in the correct physical direction. In the hands of those who understand its limitations, it has been shown to be a most valuable tool.

Simple MO theory was first applied to planar aromatic systems. Its success here—as in other types of systems—owes much to the role symmetry plays in determining the wave functions one will find. The restrictions placed on the form of the approximate solutions (resulting from use of an approximate or effective Hamiltonian) are often sufficiently strong to cause the resulting wave functions to be a reasonably good qualitative measure of the true molecular structure. Matrix elements of the Hamiltonian in the chosen basis are generally simplified to the extent that p orbitals on adjacent bonding sites (nearest neighbors) yield nonzero integrals that are the same for all pairs on similar types of atoms, while non-nearest-neighbor interactions are neglected. Diagonal elements are given another value, again depending only on the atom type involved. Thus, for an all-carbon π system

$$\begin{aligned} \langle \phi_i | \hat{\mathscr{H}} | \phi_j \rangle &= \beta \qquad i, j \text{ bonded neighbors} \\ &= \alpha \qquad i = j \\ &= 0 \qquad i, j \text{ not neighbors} \end{aligned} \tag{15.1}$$

Suitable additional refinements can be introduced to account for finite non-nearest-neighbor interactions, the presence of heteroatoms, and other effects. It is usually true, however, that the return on the investment rapidly dwindles when trying to refine a model theory such as this beyond its intrinsic capabilities.

The theory can also treat σ-systems or systems in which the σ–π separability is unavailable (nonplanar systems). Hoffmann's extended Hückel theory is a much more sophisticated model in which orbital overlap is used as a measure of the β-type of integral. It can treat mixed or σ-systems as well as π-systems and is useful in the geometrical considerations of molecules since the parameterization—and, thus, the energies—depend upon atomic separation and relative orientation.

However, we are interested more in the effect and use of group theory on such calculations and less on the quantum mechanical approximations needed to achieve practical results and, therefore, will look at some simple systems which are readily transparent in the simple Hückel approach. The application of group theoretical techniques and the physical consequences of symmetry are in no way affected by the quantum mechanical model.

We shall consider first a very simple π-system; it is very easy to treat, and our example will let us make several additional points not directly tied to the molecular symmetry or group theory.

The example is *cis*-butadiene whose π-system is shown in Figure 1 along with the symmetry elements appropriate to the C_{2v} point group to which the molecule belongs. The choice of coordinate axes agrees with the convention mentioned in Chapter 9. Although only the carbon p_z orbitals are shown, the equilibrium geometry of the entire molecule is C_{2v}, and it is this full molecular symmetry that we should use. We may state

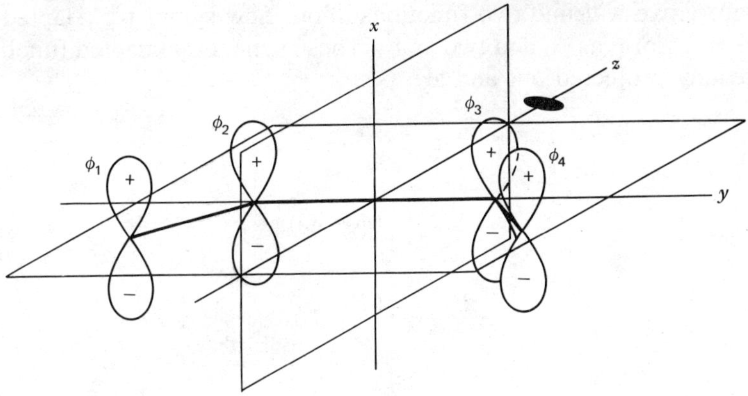

Figure 1

our purpose as finding the molecular orbitals and predicting the allowed spectral transitions arising from electronic transitions within the manifold of π-states.

The basis of p_z orbitals $\{\phi_1, \phi_2, \phi_3, \phi_4\}$ are transformed into one another by the operations of the group and, thus, will form a basis for a representation of C_{2v}. We know that the eigenfunctions will form bases for *irreducible* representations of the group so that our first step is to find a symmetry adapted basis. The usual table employed is shown in Table 1.

Table 1

C_{2v}	E	C_2	$\sigma_v^{(xz)}$	$\sigma_v^{(yz)}$		
A_1	1	1	1	1	z	x^2, y^2, z^2
A_2	1	1	−1	−1	R_z	xy
B_1	1	−1	1	−1	x, R_y	xz
B_2	1	−1	−1	1	y, R_x	yz
ϕ_1	ϕ_1	$-\phi_4$	ϕ_4	$-\phi_1$		
ϕ_2	ϕ_2	$-\phi_3$	ϕ_3	$-\phi_2$		
ϕ_3	ϕ_3	$-\phi_2$	ϕ_2	$-\phi_3$		
ϕ_4	ϕ_4	$-\phi_1$	ϕ_1	$-\phi_4$		
χ	4	0	0	−4		

Note that the phases of the p_z orbitals are important in determining the transformation properties. The character of the reducible representation Γ shows that Γ is decomposable into

$$\Gamma = 2A_2 + 2B_1 \tag{15.2}$$

Therefore, we *will* find two functions in our new symmetry adapted set which transform as A_2 and two as B_1. These symmetry adapted functions are readily projected out and are

$$\begin{aligned} A_2&: \frac{1}{\sqrt{2}}(\phi_1 - \phi_4) \\ &\frac{1}{\sqrt{2}}(\phi_2 - \phi_3) \\ B_1&: \frac{1}{\sqrt{2}}(\phi_1 + \phi_4) \\ &\frac{1}{\sqrt{2}}(\phi_2 + \phi_3) \end{aligned} \tag{15.3}$$

where the functions have been normalized neglecting orbital overlap (another simple Hückel approximation usually employed).

Without any further quantum mechanical calculations we can now state the optical selection rules. Since we are performing a 1-electron type of calculation, each single orbital product (of four orbitals for the four π electrons in butadiene) will correspond to a model molecular state. In first order, the perturbing effect of an electromagnetic field will connect (via the x, y, and z transition moment integrals) only those configurations differing in the orbital occupation of a single orbital. Since every MO in this problem is A_2 or B_1 the only possible types of transitions are $A_2 \rightarrow A_2$, $B_1 \rightarrow B_1$, $A_2 \rightarrow B_1$ (and, of course, $B_1 \rightarrow A_2$). The operators involved in the transition moment matrix elements are real (x, y, z), and the representations are real so that the allowed transitions may be specified as

z-polarized:	$A_2 \rightarrow A_2$
	$B_1 \rightarrow B_1$
y-polarized:	$A_2 \rightleftharpoons B_1$
x-polarized:	none

The simple MO solutions are listed below:

Wave function	*Symmetry*	*Energy*
$\psi_4 = 0.37\phi_1 - 0.61\phi_2 + 0.61\phi_3 - 0.37\phi_4$	A_2	-1.62β
$\psi_3 = 0.61\phi_1 - 0.37\phi_2 - 0.37 + 0.61\phi_4$	B_1	-0.62β
$\psi_2 = 0.61\phi_1 + 0.37\phi_2 - 0.37\phi_3 - 0.61\phi_4$	A_2	0.62β
$\psi_1 = 0.37\phi_1 + 0.61\phi_2 + 0.61\phi_3 + 0.37\phi_4$	B_1	1.62β

The MO's are normalized and are expressed in terms of the ϕ_i rather than the symmetry adapted basis since we often wish to examine wave functions in terms of their atomic orbital (localized) composition. Note how we can quickly confirm (or assign) the symmetry by inspection of the MO's. For example, ψ_1, the lowest energy MO of energy 1.62β (β is required physically to be negative, as is the parameter α), has coefficients for ϕ_1 and ϕ_4 that are identical in magnitude *and* sign, indicating a B_1 state; consideration of the ϕ_2 and ϕ_3 coefficients would have revealed the same result. The next higher energy state (ψ_2, $E = 0.62\beta$) is seen to be A_2 in character since ϕ_2 and ϕ_3 (for example) have equal magnitudes but opposite signs.

In the single particle approach these two lowest energy states (ψ_1 and ψ_2) would be doubly occupied. If we consider the allowed single particle excitation using our previously derived orbital selection rules and employing lower case symmetry designations for the orbitals we would obtain

Configurational change	*Symmetry selection rule*	*Polarization*	ΔE
$(b_1)^2(a_2)^2 \rightarrow (b_1)^2(a_2)(b_1')$	$A_2 \rightarrow B_1$	y	-1.24β
$(b_1)^2(a_2)^2 \rightarrow (b_1)^2(a_2)(a_2')$	$A_2 \rightarrow A_2$	z	-2.24β
$(b_1)^2(a_2)^2 \rightarrow (b_1)(a_2)^2(b_1')$	$B_1 \rightarrow B_1$	z	-2.24β
$(b_1)^2(a_2)^2 \rightarrow (b_1)(a_2)^2(a_2')$	$B_1 \rightarrow A_2$	y	-3.24β

The $A_2 \rightarrow A_2$ and $B_1 \rightarrow B_1$ transitions cannot be distinguished in this calculation since they have identical polarizations and energies. If we were to include next-nearest-neighbor interactions ($\phi_1 - \phi_3$, $\phi_2 - \phi_4$) it is readily shown that ψ_1 and ψ_4 are lowered in energy while ψ_2 and ψ_3 are raised. This results in a shift to higher energies (blue shift) of the $B_1 \rightarrow B_1$ transition and a shift to lower energy (red shift) of the $A_2 \rightarrow A_2$ transition. Thus, we would really expect to be able to resolve the two transitions in this way.

The addition of the 1–3 and 2–4 next-nearest-neighbor interactions in Figure 2 in no way affects our symmetry considerations or our symmetry classification. We have used the full proper symmetry of the molecule; only quantitative changes are employed in keeping certain numerical terms rather than neglecting them. This raises an interesting point, however. So far as the quantum mechanical nearest-neighbor calculation is concerned, *any* planar and topologically linear arrangement of the p_z orbitals will yield the same eigenvalues and eigenfunctions since the mathematics will be unaltered. Some of the symbolically represented structures permitted are seen in Figure 2. Note that *trans*-butadiene is

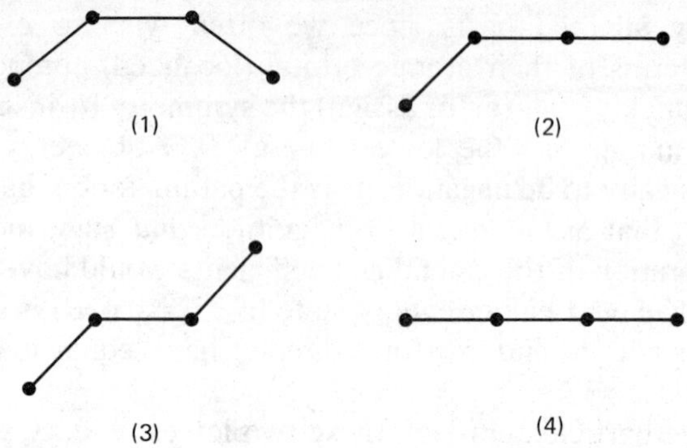

Figure 2

represented; this means that although they have different molecular symmetries (C_{2h} for the *trans* case as opposed to C_{2v} for the *cis* case), *cis*- and *trans*-butadiene will—in the simple Hückel calculation—have identical energies and wave functions. The labeling of the states will differ, but the physical effects will be the same. This fact remains unaltered even when next-nearest-neighbor interactions are included, provided, of course, that the 1–3 and 2–4 distances and orientations are maintained equal. Only when we get to the inclusion of the 1–4 interaction will a quantitative difference be introduced. We see then that in certain types of model calculations where all interactions are not included we may be able to operate with an effective symmetry different than the true one. It may be advantageous to work with the alternate symmetry classification in our quantum mechanical manipulations although we should ultimately return to the proper labeling and symmetry classification; we do have to compare our calculations with experiment and while mere mortals often have to resort to approximate treatments, mother nature is very good at displaying rigorous results.

If we are permitted by our model to use a lower symmetry this may ease the manipulations we have to perform in terms of the classification of states and transitions and so forth. We will not lose any calculated degeneracies although *essential* degeneracies in the proper group may become *accidental* in the lower symmetry classification. On the other hand, the ability to use a group of higher symmetry may allow us to predict degeneracies which, again, would ultimately show up in the calculation but would not be classified as essential before the fact. In such an instance we could well expect a more rigorous calculation demanding the use of the proper group to show *near* degeneracies.

Problems

15.1 By inspection of the π wave functions for *cis*-butadiene, write down their symmetry classification when used in the simple Hückel treatment of *trans*-butadiene.

15.2 Using the results of problem 11.6, find the π-MO's of benzene as a D_{6h} molecule. Repeat the treatment in the reduced symmetry obtained by lowering the hydrogens uniformly out of the carbon atom plane and moving each uniformly clockwise with respect to an axis through the center of the undistorted molecule and perpendicular to the undistorted molecular plane.

Problem 15.2 shows that it is much easier to write down the benzene π-MO's using C_6 symmetry rather than D_{6h}. Actually, the use of C_N-type symmetry is most important in solids where we have translational as well as point group symmetry elements. Our point can be made most simply by considering a one-dimensional array of N centers with translational symmetry relating neighboring centers. The physical argument is made that since N is very large (approximately 10^{20}) the basic properties of the crystal will be determined by the internal geometry and not by edge effects; that is, if we were to connect one end of the crystal with the other there would be essentially no physical effect. This allows us to employ cyclic boundary conditions; it is topologically equivalent to viewing the crystal as a cycle of N units and this allows us to use C_N symmetry. The resulting 1-electron functions are the Bloch functions and, as far as their symmetry properties are concerned, are nothing more than MO functions for a very large cyclic polyene. All three dimensions of a real crystal can be handled in this way as well as additional symmetry within the unit cell. This approach forms the basic use of symmetry and group theory in the band structure of solids.

Problem

15.3 Consider a one-dimensional lattice of N unit cells, each cell containing two centers labeled a and b with associated 1-electron localized orbitals $\phi_{n,a}$ and $\phi_{n,b}$, n specifying the particular unit cell $(1 \leq n \leq N)$ in which the site is located. Each $\phi_{n,a}$ is related to $\phi_{n+1,a}$ by translational symmetry, as are $\phi_{n,b}$ and $\phi_{n+1,b}$. Employ cyclic boundary conditions and set up the 1-electron quantum mechanical problem. Assume MO parameters $\alpha_a \neq \alpha_b$ and nearest-neighbor-only β's (there are *two* β's in this problem) and exhibit the energy band structure that results.

The examples we have worked with to illustrate group theory applied to molecular orbital calculations have involved π-systems. We have done so mainly because π-MO theory is generally known to most people and because it is particularly easy to visualize. It should be clear by now, hopefully, that the approach is general and can be applied to any calculation on any type of molecule in which one wants to initiate the calculation with symmetry adapted orbitals.

VALENCE BOND THEORY AND DIRECTED VALENCE; HYBRIDIZATION

A principal flaw in MO theory is the underemphasis on electron correlation. Molecular orbitals are doubly occupied (by spin-paired electrons) which allows electrons to coexist in the same region of space and which allows a large repulsive coulomb interaction. Thus, the simple MO wave function for the ground singlet state of H_2 employing the restricted basis of a $1s$ orbital on each nucleus (s_a, s_b) is

$$\begin{aligned}\psi &= \mathcal{A}\left(\frac{s_a + s_b}{\sqrt{2+2S}}\right)\left(\frac{s_a + s_b}{\sqrt{2+2S}}\right)\alpha\beta \\ &= (2+2S)^{-1}\{s_a s_a + s_b s_b + s_a s_b + s_b s_a\}\frac{(\alpha\beta - \beta\alpha)}{\sqrt{2}}\end{aligned} \tag{15.4}$$

where $S = \langle s_a \mid s_b \rangle$ is the appropriate overlap integral. The wave function multiplied out in terms of the atomic orbitals involved shows that terms like $s_a s_a + s_b s_b$—the "ionic" terms—have equal billing with $s_a s_b + s_b s_a$—the "covalent" terms. There is little effort to correlate the motions of the electrons, and, indeed, the MO wave functions assume an independence to their motion that is, of course, not really present.

The valence bond approach in its simplest application is basically to remove ionic terms. Although it tends to overdo it, it does introduce correlation by requiring bonding electrons to reside in different spatial orbitals. Accordingly, for H_2 the VB approximation would be

$$\begin{aligned}\psi &= \mathcal{A} s_a s_b \frac{\alpha\beta - \beta\alpha}{\sqrt{2}}(1+S^2)^{-1/2} \\ &= (2+2S^2)^{-1/2}\{s_a s_b + s_b s_a\}\frac{(\alpha\beta - \beta\alpha)}{\sqrt{2}}\end{aligned} \tag{15.5}$$

using the same atomic orbital basis as in the MO case.

Generally in the MO approach the use of simple hydrogen-like or hybrid orbitals will not affect the final wave function. A molecular orbital made up of a linear combination of localized orbitals can be written in an infinite variety of ways as a sum over equivalent bases (bases related by a similarity transformation) but it remains the same molecular orbital. In valence bond theory, however, the electrons are selectively placed in orbitals which hopefully maximize the bonding power. Lone pairs and inner core electrons may doubly occupy the same spatial orbital but the bonding electrons do not and enter the VB wave function in the fashion of the covalent H_2 example above. The stability of the system arises from these bonding pairs.

Although being in different spatial orbitals is the characterizing feature by which the VB method introduces correlation, still the bonding orbitals must have favorable overlap in order to provide the optimum stabilization which arises basically through the 1-electron exchange integrals. In short, the form of the atomic basis set *will* make a difference in the end result; we want to choose bonding orbitals localized on the bonded atoms that overlap well with each other. Thus, hybridization is very important in the valence bond picture.

When symmetry is present in a molecule, the orbitals employed may well be related by the group elements. As the following examples will show, the transformation properties of the irreducible basis sets provide the key in forming the desired equivalent hybrids. Consider first the water molecule (in the Bonn–Oppenheimer or rigid nuclei approximation) as depicted in Figure 3. Suppose we wish to form directional hybrids out of the oxygen $2s$, $2p_x$, $2p_y$, and $2p_z$ atomic orbitals that have maximum overlap with the two hydrogen atoms. We are interested in finding the transformation that will take us from the "pure" s and p basis to a basis of hybrid such that two of the resulting new basis functions will have directional character and will be directed at the hydrogen atoms H_1 and

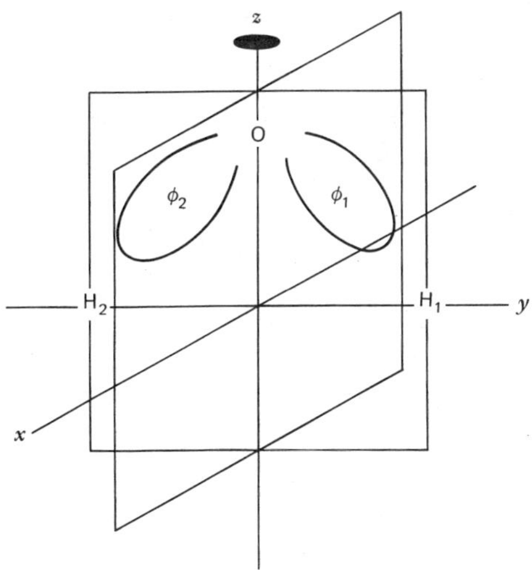

Figure 3

H_2. We shall demand that the new basis of oxygen functions also be orthogonal and that, since symmetry dictates it, the directed hybrids ϕ_1 and ϕ_2 depicted schematically in Figure 3 be equivalent. The equivalence of ϕ_1 and ϕ_2 means that they are related by the group elements as shown in the lower portion of the character table (Table 2); the fact that the two bases are orthogonal means that the similarity transformation matrix will be unitary.

Table 2

C_{2v}	E	C_2	$\sigma_v^{(xz)}$	$\sigma_v^{(yz)}$	
A_1	1	1	1	1	z, x^2, y^2, z^2
A_2	1	1	−1	−1	xy
B_1	1	−1	1	−1	x, xz
B_2	1	−1	−1	1	y, yz
ϕ_1	ϕ_1	ϕ_2	ϕ_2	ϕ_1	
ϕ_2	ϕ_2	ϕ_1	ϕ_1	ϕ_2	
Γ_ϕ	2	0	0	2	

As is well known, the real form of the hydrogen-like orbitals have transformational properties like the various polynomials of x, y, z; thus, s-states transform like $x^0 = y^0 = z^0 = 1$ (a radial dependence only), p-states like x, y, and z, and for the d-states xy, xz, yz, z^2, and $x^2 - y^2$. Our character table then shows us that the oxygen $2s$ and $2p_z$ orbitals transform as A_1, p_x as B_1 and p_y as B_2.

We begin by transforming the equivalent orbitals ϕ_1 and ϕ_2 into symmetry adapted orbitals. $\Gamma_\phi = \{\phi_1, \phi_2\}$ is a basis for a reducible representation and is decomposed into $\Gamma_\phi = A_1 + B_2$. Applying the proper projection operators we obtain the (properly normalized; recall that the AO's are orthonormal) symmetry adapted functions:

$$\begin{aligned} A_1&: \psi_{A_1} = \frac{1}{\sqrt{2}}(\phi_1 + \phi_2) \\ B_2&: \psi_{B_2} = \frac{1}{\sqrt{2}}(\phi_1 - \phi_2) \end{aligned} \tag{15.6}$$

The two sets are related by the unitary transformation U:

$$(\phi_1, \phi_2)\begin{pmatrix} \frac{1}{\sqrt{2}} & \frac{1}{\sqrt{2}} \\ \frac{1}{\sqrt{2}} & -\frac{1}{\sqrt{2}} \end{pmatrix} = (\psi_{A_1}, \psi_{B_2}) \tag{15.7}$$

Of more significance is the inverse relation $\phi = \psi \mathbf{U}^\dagger$

$$(\phi_1, \phi_2) = (\psi_{A_1}, \psi_{B_2}) \begin{pmatrix} \frac{1}{\sqrt{2}} & \frac{1}{\sqrt{2}} \\ \frac{1}{\sqrt{2}} & -\frac{1}{\sqrt{2}} \end{pmatrix} \tag{15.8}$$

for it shows us that given *any* two functions that transform as A_1 and B_2 the above transformation will allow us to form equivalent orbitals having—by definition—the symmetry transformation properties of ϕ_1 and ϕ_2. Thus, we could transform the oxygen orbital pairs $2s$, $2p_y$ or $2p_z$, $2p_y$ to form an equivalent ϕ-pair; or, since any combination of $2s$ and $2p_z$ transforms as A_1, we could use the pair $\{c_s 2s + c_z 2p_z, 2p_y\}$, where $c_s 2s + c_z 2p_z$ is an arbitrary but normalized mixture. The normalization requirement $c_s^2 + c_z^2 = 1$ is one constraint; the other requirement we need is the orthogonality between ϕ_1 and ϕ_2 which is automatically satisfied. The one degree of freedom remaining, then, is used to guarantee maximum overlap with the hydrogen bonding functions which, in this case, is done by pointing the oxygen hybrids at the appropriate hydrogen atom.

Problems

15.4 p-functions act like vectors in the sense that the normalized linear combination $\alpha p_x + \beta p_y$ is again a p-orbital which is largest in a direction such that $\tan\theta = \beta/\alpha$, where θ is the angle between this direction and the x-axis. Find the value of c_p (and, thus, c_s) that leads to an angle of 105° between the equivalent hybrids in the water molecule just discussed.

15.5 Having found in problem 15.4 two functions in the new (hybrid) basis, there are yet two more to be specified. What contraints remain for these last two (not necessarily equivalent) functions and what possible sets of functions could one produce?

15.6 If d-orbitals (for oxygen, the $3d$ function) were also considered in finding the *bonding* hybrids in water, what would the most general form of the hybrids be as dictated by the symmetry requirements?

15.7 Consider the trigonal pyramid ammonia molecule characterized by C_{3v} symmetry. Find the transformation matrix that can be used to construct equivalent localized directional hybrids out of the nitrogen $2s$ and $2p$ orbitals having maximum overlap with the three hydrogen atoms, and show that all four functions of the new basis set are completely determined by the symmetry requirements and the detailed molecular geometry specified by HNH bond angles of 107°.

15.8 Find the appropriate linear combinations of *s* and *p* orbitals for an atom that describe *sp* (linear), sp^2 (trigonal), and sp^3 (tetrahedral) hybridization. As models, choose linear CH_2, trigonal planar CH_3, and tetrahedral CH_4. Employ the simplest possible point group applicable to the symmetry of the desired equivalent orbitals.

The final considerations in the water or ammonia molecule problems are mostly quantum mechanical in nature rather than group theoretical. Our main point has been to see how the group character table allows us to construct the hybrids which a particular quantum mechanical approximation—the valence bond method—deems desirable. The final many-electron wave function must, of course, have a definite symmetry and we would approach this problem just as we did in Chapter 14 and other earlier chapters.

SOLUTIONS TO PROBLEMS

15.1 *Trans*-butadiene has the symmetry C_{2h} as shown in Figure 4 and the character table.

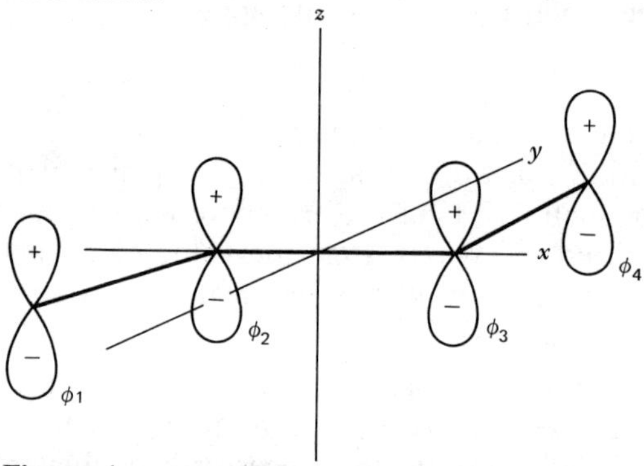

Figure 4

C_{2h}	E	C_2	i	σ_h	
A_g	1	1	1	1	
B_g	1	−1	1	−1	
A_u	1	1	−1	−1	z
B_u	1	−1	−1	1	x, y

Inspection of the wave functions listed in the text for *cis*-butadiene reveals the following properties under the C_{2h} operations

	C_2	i	σ_h
ψ_1, ψ_3	+	−	−
ψ_2, ψ_4	−	+	−

Thus, ψ_1 and ψ_3 in *trans*-butadiene are A_u states while ψ_2 and ψ_4 are B_g. For *trans*-butadiene the orbital optical selection rules will be

$A_u \rightarrow A_u$	forbidden
$B_g \rightarrow B_g$	forbidden
$A_u \rightleftharpoons B_g$	allowed for x or y polarization

Thus, the predicted spectra for the *cis* and *trans* cases differ significantly because of their different symmetry and suggest that we could distinguish between them on this basis.

15.2 The solution to problem 11.6 shows that the representation space spanned by the six p_z benzene orbitals may be decomposed as

$$\Gamma = B_{2g} + E_{1g} + A_{2u} + E_{2u}$$

so that we expect two sets of doubly degenerate pairs (the E_{1g} and E_{2u} sets) and two sets of singly degenerate functions (the B_{2g} and A_{2u} functions). Using the projection operator technique we find for B_{2g} and A_{2u}

$$B_{2g}: \frac{1}{\sqrt{6}}(\phi_1 - \phi_2 + \phi_3 - \phi_4 + \phi_5 - \phi_6)$$

$$A_{2u}: \frac{1}{\sqrt{6}}(\phi_1 + \phi_2 + \phi_3 + \phi_4 + \phi_5 + \phi_6)$$

For E_{1g} we form

$$\frac{1}{\sqrt{12}}(2\phi_1 + \phi_2 - \phi_3 - 2\phi_4 - \phi_5 + \phi_6) \qquad (1)$$

$$\frac{1}{\sqrt{12}}(\phi_1 + 2\phi_2 + \phi_3 - \phi_4 - 2\phi_5 - \phi_6) \qquad (2)$$

$$\frac{1}{\sqrt{12}}(-\phi_1 + \phi_2 + 2\phi_3 + \phi_4 - \phi_5 - 2\phi_6) \qquad (3)$$

which are not linearly independent ($(1)-(2)+(3)=0$) since we know we can have only two independent functions in an E representation. We could choose (1) and (2), or the set $(1)+(2)$ and (3) (often chosen in practice)

$$\tfrac{1}{2}(\phi_1 + \phi_2 - \phi_4 - \phi_5) \qquad (1')$$

$$\frac{1}{\sqrt{12}}(-\phi_1 + \phi_2 + 2\phi_3 + \phi_4 - \phi_5 - 2\phi_6) \qquad (2')$$

or the "symmetric" set (1)+(2) and (2)+(3):

$$\tfrac{1}{2}(\phi_1+\phi_2-\phi_4-\phi_5) \qquad (1'')$$

$$\tfrac{1}{2}(\phi_2+\phi_3-\phi_5-\phi_6) \qquad (2'')$$

For E_{2u} we obtain the linearly dependent set

$$\frac{1}{\sqrt{12}}(2\phi_1-\phi_2-\phi_3+2\phi_4-\phi_5-\phi_6) \qquad (1)$$

$$\frac{1}{\sqrt{12}}(-\phi_1+2\phi_2-\phi_3-\phi_4+2\phi_5-\phi_6) \qquad (2)$$

$$\frac{1}{\sqrt{12}}(-\phi_1-\phi_2+2\phi_3-\phi_4-\phi_5+2\phi_6) \qquad (3)$$

from which we will choose (for reasons to be seen below) the linearly independent pair (1)−(2) and (3)

$$\tfrac{1}{2}(\phi_1-\phi_2+\phi_4-\phi_5) \qquad (1')$$

$$\frac{1}{\sqrt{12}}(-\phi_1-\phi_2+2\phi_3-\phi_4-\phi_5+2\phi_6) \qquad (2')$$

If the molecule is now distorted (see Figure 5) the symmetry is reduced to C_6 although, clearly, if we do not intend to modify the π-orbital matrix elements we will not affect the results of our calculation. We recall the general simplicity of the C_N group and its character table and find the decomposition of Γ now to be

$$\Gamma=\sum_{n=0}^{N-1}\Gamma_n$$

$$\chi_n(C_N^m)=(e^{i(2\pi/N)n})^m$$

Each one-dimensional representation occurs just once and the projected wave functions will be

$$\Gamma_n:\ \psi_n=\sum_{m=1}^{N}(e^{i(2\pi/N)n})^m\phi_m$$

We recall also that although C_N consists of only one-dimensional

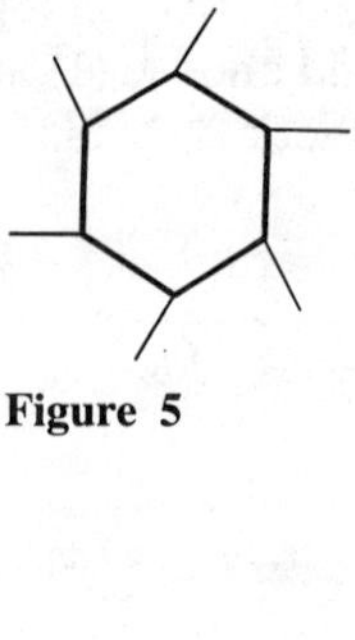

Figure 5

representations if the quantum mechanical Hamiltonian is real—as we presume to be in the present situation—that the Γ_n and $\Gamma_{N-n} = \Gamma_n^*$ representations will be degenerate.

For our case of $N = 6$ we have for $n = 0$ and $n = 3$ the following unique functions:

	Γ_n of C_6	Γ of D_{6h}
$n = 0;$	$\psi_0 = \frac{1}{\sqrt{6}}(\phi_1 + \phi_2 + \phi_3 + \phi_4 + \phi_5 + \phi_6)$	A_{2u}
$n = 3;$	$\psi_3 = \frac{1}{\sqrt{6}}(-\phi_1 + \phi_2 - \phi_3 + \phi_4 - \phi_5 + \phi_6)$	B_{2g}

which to a sign agrees—as they must—with the unique A_{2u} and B_{2g} functions of D_{6h}. The other (really degenerate) representations and functions occur as complex conjugates; we may readily write them out in complex form or, in order to compare them to the D_{6h} function, we can form the real sum and difference functions.

$$\psi_n^+ \propto (\psi_n + \psi_{N-n}) = (\psi_n + \psi_n^*) = 2\,\mathrm{Re}\,\psi_n$$

$$\psi_n^- \propto \frac{1}{i}(\psi_n - \psi_{N-n}) = \frac{1}{i}(\psi_n - \psi_n^*) = 2\,\mathrm{Im}\,\psi_n$$

Thus, since $\cos \pi/3 = \frac{1}{2}$, $\sin \pi/3 = \sqrt{3}/2$,

$$n = 1 \qquad \psi_1^+ = \frac{1}{\sqrt{12}}(\phi_1 - \phi_2 - 2\phi_3 - \phi_4 + \phi_5 + 2\phi_6)$$

$$\psi_1^- = \tfrac{1}{2}(\phi_1 + \phi_2 - \phi_4 - \phi_5)$$

which are clearly the E_{1g} functions; while for $n = 2$,

$$n = 2 \qquad \psi_2^+ = \frac{1}{\sqrt{12}}(-\phi_1 - \phi_2 + 2\phi_3 - \phi_4 - \phi_5 + 2\phi_6)$$

$$\psi_2^- = \tfrac{1}{2}(\phi_1 - \phi_2 + \phi_4 - \phi_5)$$

which are clearly the E_{2u} functions.

The point of these latter manipulations is to show that it is much simpler in this case to work with the group of lower symmetry (C_6 versus D_{6h}) in forming the symmetry adapted functions; they can be written down essentially by inspection.

15.3 Each $\phi_{a,n}$ and $\phi_{b,n}$ are related to those in neighboring cells by C_N operations so that for each type we can project out the symmetry adapted sets:

$$\psi_{a,k} = \frac{1}{\sqrt{N}} \sum_n e^{ikn} \phi_{n,a}$$

$$\psi_{b,k} = \frac{1}{\sqrt{N}} \sum_n e^{ikn} \phi_{n,b}$$

$$k = \frac{2\pi}{N} m, \qquad m = 0, 1, 2, \ldots, N-1$$

where we have assumed the localized functions to be orthonormal. For a given k there are two symmetry functions, so we must diagonalize for each k subset. Using the integral restrictions,

$$\langle \phi_{n,a} | \mathscr{H} | \phi_{n',a'} \rangle = \alpha_a \delta_{nn'}$$

$$\langle \phi_{n,b} | \hat{\mathscr{H}} | \phi_{n',b'} \rangle = \alpha_b \delta_{nn'}$$

$$\langle \phi_{n,a} | \hat{\mathscr{H}} | \phi_{n-1,b} \rangle = \beta_2$$

$$\langle \phi_{n,a} | \hat{\mathscr{H}} | \phi_{n,b} \rangle = \beta_1$$

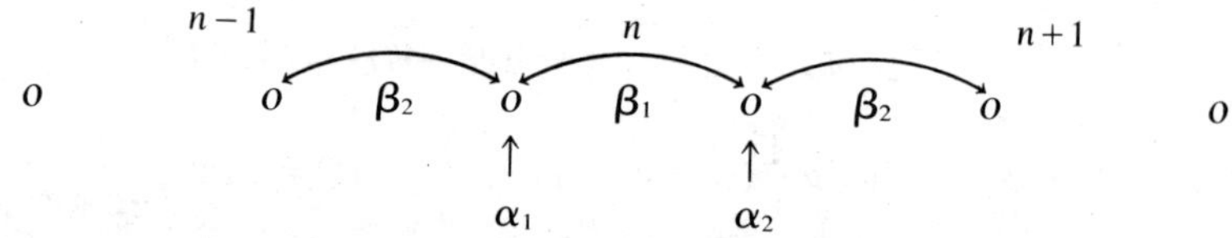

we find

$$\langle \psi_{a,k} | \hat{\mathscr{H}} | \psi_{a,k} \rangle = \alpha_1$$

$$\langle \psi_{b,k} | \hat{\mathscr{H}} | \psi_{b,k} \rangle = \alpha_2$$

$$\langle \psi_{a,k} | \hat{\mathscr{H}} | \psi_{b,k} \rangle = \beta_1 + \beta_2 e^{-ik}$$

so that the secular determinant becomes

$$\begin{vmatrix} \alpha_1 - \varepsilon_k & \beta_1 + \beta_2 e^{-ik} \\ \beta_1 + \beta_2 e^{+ik} & \alpha_2 - \varepsilon_k \end{vmatrix} = 0$$

with solutions

$$2E_k^{\pm} = \alpha_1 + \alpha_2 \pm [(\alpha_1 - \alpha_2)^2 + 4(\beta_1^2 + \beta_2^2 + 2\beta_1\beta_2 \cos k)]^{1/2}$$

Consider all the α and β parameters to be negative. Then we see that as k varies from 0 to 2π, E_k^+, and E_k^- vary smoothly in energy, each forming a band of energies. The minimum energy for E_k^+ (the bottom of the upper band) is

$$2E_{k,\min}^{+} = 2E_{k=\pi}^{+} = \alpha_1 + \alpha_2 + [(\alpha_1 - \alpha_2)^2 + 4(\beta_1 - \beta_2)^2]^{1/2}$$

while the maximum energy for E_k^- (the top of the lower band) is

$$2E_{k,\max}^{-} = 2E_{k=\pi}^{-} = \alpha_1 + \alpha_2 - [(\alpha_1 - \alpha_2)^2 + 4(\beta_1 - \beta_2)^2]^{1/2}$$

In general, $E^{+}_{k,\min} > E^{-}_{k,\max}$ and an energy gap exists. This gap in the overall band structure has come in because of the additional structure within the unit cell (two sites instead of just one). In general, the gap

$$\Delta E = [(\alpha_1 - \alpha_2)^2 + 4(\beta_1 - \beta_2)^2]^{1/2}$$

will only vanish if $\alpha_1 = \alpha_2$ and $\beta_1 = \beta_2$, in which case we have a simple cyclic lattice of $2N$ equivalent centers. The band structure becomes more complicated as more dimensions and more internal structure are considered but the basic symmetry considerations are the same.

15.4 The orbitals involved are

$$\phi_1 = \frac{1}{\sqrt{2}}(c_s s + c_z p_z) + \frac{1}{\sqrt{2}} p_y$$

$$\phi_2 = \frac{1}{\sqrt{2}}(c_s s + c_z p_z) - \frac{1}{\sqrt{2}} p_y$$

We can choose $c_s > 0$ and $c_p < 0$ in order to have positive labeled functions coinciding with the pictorial representation of ϕ_1 and ϕ_2 in the text. Both ϕ_1 and ϕ_2 make an angle Θ ($0 \leq \Theta \leq 90°$) with the minus z-axis such that

$$\tan \Theta = \frac{1}{-c_z}$$

$$\Theta = \tfrac{1}{2}(105°)$$

$$c_z = -0.767$$

so that

$$c_s = (1 - c_z^2)^{1/2} = +0.641$$

We can use a more general approach [see, for example, G. Kimball, *Ann. Rev. Phys. Chem.* **2,** 177 (1951)]. ϕ_1 and ϕ_2 can be interpreted as sp^n hybrides of the general form

$$\phi_{(n)} = \frac{s + \sqrt{n}p}{(1+n)^{1/2}}$$

Two such hybrides pointing in directions differing by the angle η can be shown to have an overlap integral

$$\langle \phi_{(n_1)} | \phi_{(n_2)} \rangle = [(1+n_1)(1+n_2)]^{-1/2}(1 + (n_1 n_2)^{1/2} \cos \eta)$$

If they are to be orthogonal

$$1 + (n_1 n_2)^{1/2} \cos \eta = 0$$

and if, in addition, they are equivalent ($n_1 = n_2$)

$$1 + n \cos \eta = 0$$

In our case above, the orbitals are equivalent and orthogonal so that since $\eta = 105°$, $n = 3.864$. The total weight of p orbitals in the hybrid is $n/(1+n)$ and in this case is given by

$$\frac{n}{1+n} = \tfrac{1}{2}(c_z^2 + 1)$$

This yield $c_z = -0.767$ (by our choice of sign for c_p) and then $c_s = +0.641$, as before.

15.5 One of the remaining two symmetry adapted orbitals *on oxygen* is p_x which transforms as B_1; the other is the remaining combination of s and p_z orthogonal to that combination $c_s s + c_z p_z$ used in preparing the equivalent ϕ-hybrids, namely,

$$-c_z s + c_s p_z = 0.767\ s + 0.641\ p_z$$

an $sp^{0.698}$ hybrid which, of course, also transforms as A_1. We can take these two separately or could combine them; we could, if we like, require that they form a pair of equivalent hybrid orbits in the x–z-plane and proceed just as in the case of the ϕ_1, ϕ_2 hybrids. The hybrid mixture of s and p_z is already determined in this case so that the interorbital angle $2\theta'$ is also predetermined

$$\tan \theta' = \frac{1}{0.641}$$

The interorbital angle is

$$2\Theta' = 114.7°$$

These last two equivalent hybrids correspond to the type $sp^{2.398}$ since here

$$n = \frac{(0.641)^2 + (1)^2}{(0.767)^2} = 2.398$$

15.6 The symmetry mixture is still the same, namely,

$$\phi_{\pm} = \frac{1}{\sqrt{2}}(\psi_{A_1} \pm \psi_{B_2})$$

where now, allowing for possible d-orbital contributions to both ψ_{A_1} and ψ_{B_2}, we can have the general forms

$$\psi_{A_1} = c_s s + c_z p_z + c_{z^2} d_{z^2} + c_{x^2-y^2} d_{x^2-y^2}$$

$$\psi_{B_2} = c_y p_y + c_{yz} d_{yz}$$

Inclusion of the d-orbitals makes the geometrical considerations more complicated (see Kimball reference in problem 15.4).

15.7 See Figure 6 and its character table (Table 3).

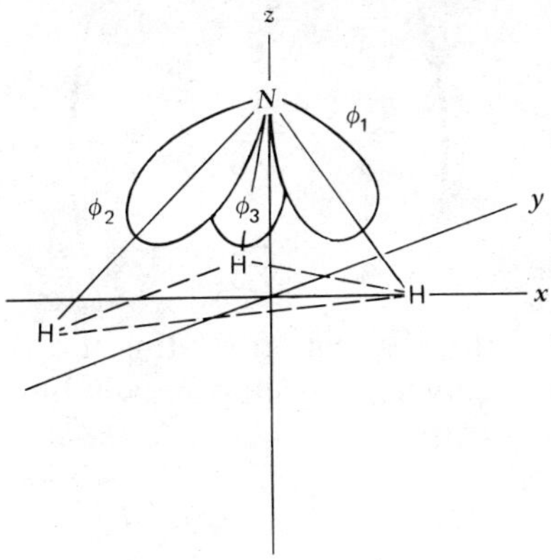

Figure 6

Table 3

C_{3v}	E	$2C_3$		$3\sigma_v$			
A_1	1	1			1		z, s
A_2	1	1			-1		
E	2	-1			0		x, y
		C_3	$\bar{C}_3$	σ_1	σ_2	σ_3	
ϕ_1	ϕ_1	ϕ_2	ϕ_3	ϕ_1	ϕ_3	ϕ_2	
ϕ_2	ϕ_2	ϕ_3	ϕ_1	ϕ_3	ϕ_2	ϕ_1	
ϕ_3	ϕ_3	ϕ_1	ϕ_2	ϕ_2	ϕ_1	ϕ_3	
Γ	3	0			1		

$$\Gamma = A_1 + E$$

We proceed exactly as in the water molecule example. Note that C_3 corresponds to a clockwise rotation looking down the z-axis and that the various σ_j each contain one of the equivalent nitrogen hybrids ϕ_j.

The nitrogen symmetry orbitals are (recall the choice we have for the E-type symmetry orbitals)

$$\psi_{A_1} = \frac{1}{\sqrt{3}}(\phi_1 + \phi_2 + \phi_3)$$

$$\psi_E^{(1)} = \frac{1}{\sqrt{6}}(2\phi_1 - \phi_2 - \phi_3)$$

$$\psi_E^{(2)} = \frac{1}{\sqrt{2}}(\phi_2 - \phi_3)$$

so that

$$(\phi_1, \phi_2, \phi_3)\begin{pmatrix} \frac{1}{\sqrt{3}} & \frac{2}{\sqrt{6}} & 0 \\ \frac{1}{\sqrt{3}} & -\frac{1}{\sqrt{6}} & \frac{1}{\sqrt{2}} \\ \frac{1}{\sqrt{3}} & -\frac{1}{\sqrt{6}} & -\frac{1}{\sqrt{2}} \end{pmatrix} = (\psi_{A_1}, \psi_E^{(1)}, \psi_E^{(2)})$$

and the general transformation relating the s, z, x, and y nitrogen orbitals to the equivalent orbitals ϕ_j will be

$$(\phi_1, \phi_2, \phi_3) = (c_s s + c_{pz} p_z, p_x, p_y)\begin{pmatrix} \frac{1}{\sqrt{3}} & \frac{1}{\sqrt{3}} & \frac{1}{\sqrt{3}} \\ \frac{2}{\sqrt{6}} & -\frac{1}{\sqrt{6}} & -\frac{1}{\sqrt{6}} \\ 0 & \frac{1}{\sqrt{2}} & -\frac{1}{\sqrt{2}} \end{pmatrix}$$

We could determine c_s and c_{p_z} by using the specified molecular geometry, working out the angle that ϕ_1 makes with the x-axis, for example, and requiring c_{p_z} to yield the proper result. It is simpler, however, to use the equivalent hybrid method illustrated in problem 15.5. From it, we conclude that for equivalent orthogonal sp^n hybrids η degrees apart,

$$1 + n \cos \eta = 0$$

or, since $\eta = 107°$ in the present case

$$n = -\frac{1}{\cos \eta} = 3.420$$

Using the orbital ϕ_1 (we could use any of them) this means that

$$\frac{n}{1+n} = \tfrac{1}{3}c_{p_z}^2 + \tfrac{2}{3}$$

$$c_s = (1 - c_{p_z}^2)^{1/2}$$

so that, choosing $c_{p_z} < 0$ to have positive overlap with the hydrogen orbitals, $c_{p_z} = -0.567$ and $c_s = 0.823$. These values determine the coefficients of the bonding hybrids. The one remaining hybrid ϕ_{nb} (representing the orbital for the nonbonding pair of electrons in ammonia) will be (orthogonal to $0.823\, s - 0.567\, p_z$)

$$\phi_{nb} = 0.567\, s + 0.823\, p_z$$

15.8 We can employ groups whose symmetry is lower than the various molecular models since we are only interested in finding the proper linear combinations. As long as we have this freedom, it may pay us to use groups that separate *x*, *y*, and *z* as much as possible. For linear CH_2 with the hydrogen atoms placed along the positive and negative *x*-axis we could use C_2 but we may also select D_2; D_2 is the simplest group which separates *x*, *y*, and *z* and so we employ it (see Figure 7).

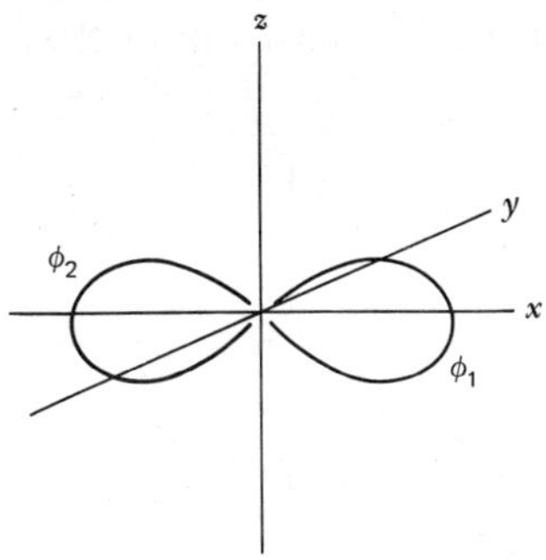

Figure 7

	E	$C_2(z)$	$C_2(y)$	$C_2(x)$
ϕ_1	ϕ_1	ϕ_2	ϕ_2	ϕ_1
ϕ_2	ϕ_2	ϕ_1	ϕ_1	ϕ_2
Γ	2	0	0	2

$$\Gamma = A + B_3$$

Thus

$$\psi_A = \frac{1}{\sqrt{2}}(\phi_1 + \phi_2)$$

$$\psi_{B_3} = \frac{1}{\sqrt{2}}(\phi_1 - \phi_2)$$

$$(\psi_A, \psi_{B_3}) = (\phi_1, \phi_2)\begin{pmatrix} \frac{1}{\sqrt{2}} & \frac{1}{\sqrt{2}} \\ \frac{1}{\sqrt{2}} & -\frac{1}{\sqrt{2}} \end{pmatrix}$$

so that we can choose

$$\phi_1 = \frac{1}{\sqrt{2}} s + \frac{1}{\sqrt{2}} p_x$$

$$\phi_2 = \frac{1}{\sqrt{2}} s - \frac{1}{\sqrt{2}} p_x$$

For trigonal planar CH_3 we take the group C_3 treating it as consisting of three one-dimensional representations recalling from problem 11.5 how $x \pm iy$ transform in this group. (See Figure 8.)

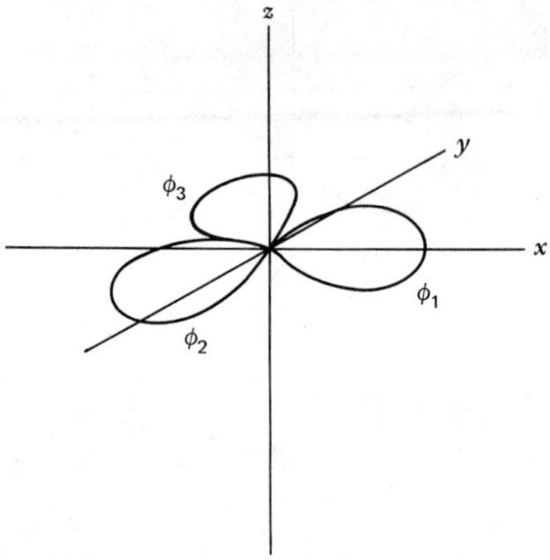

Figure 8

C_3	E	C_3	C_3^2	
Γ_1	1	1	1	s, z
Γ_2	1	ε^*	ε	$x - iy$
Γ_3	1	ε	ε^*	$x + iy$
ϕ_1	ϕ_1	ϕ_2	ϕ_3	
ϕ_2	ϕ_2	ϕ_3	ϕ_1	
ϕ_3	ϕ_3	ϕ_1	ϕ_2	
Γ	3	0	0	
		$\Gamma = \Gamma_1 + \Gamma_2 + \Gamma_3$		

$$\varepsilon = e^{2\pi i/3}$$

$$\varepsilon + \varepsilon^* = -1$$

$$\varepsilon - \varepsilon^* = i\sqrt{3}$$

$$\psi_{\Gamma_1} = \frac{1}{\sqrt{3}}(\phi_1 + \phi_2 + \phi_3)$$

$$\psi_{\Gamma_2} = \frac{1}{\sqrt{3}}(\phi_1 + \varepsilon\phi_2 + \varepsilon^*\phi_3)$$

$$\psi_{\Gamma_3} = \frac{1}{\sqrt{3}}(\phi_1 + \varepsilon^*\phi_2 + \varepsilon\phi_3)$$

$$(\psi_{\Gamma_1}, \psi_{\Gamma_2}, \psi_{\Gamma_3}) = (\phi_1, \phi_2, \phi_3) \begin{pmatrix} \frac{1}{\sqrt{3}} & \frac{1}{\sqrt{3}} & \frac{1}{\sqrt{3}} \\ \frac{1}{\sqrt{3}} & \frac{1}{\sqrt{3}}\varepsilon & \frac{1}{\sqrt{3}}\varepsilon^* \\ \frac{1}{\sqrt{3}} & \frac{1}{\sqrt{3}}\varepsilon^* & \frac{1}{\sqrt{3}}\varepsilon \end{pmatrix}$$

Remembering that the transformation matrix is unitary we find

$$\phi_1 = \frac{1}{\sqrt{3}}\left\{s + \frac{1}{\sqrt{2}}(p_x - ip_y) + \frac{1}{\sqrt{2}}(p_x + ip_y)\right\}$$

$$= \frac{1}{\sqrt{3}}(s + \sqrt{2}\,p_x)$$

$$\phi_2 = \frac{1}{\sqrt{3}}\left\{s + \frac{\varepsilon^*}{\sqrt{2}}(p_x - ip_y) + \frac{\varepsilon}{\sqrt{2}}(p_x + ip_y)\right\}$$

$$= \frac{1}{\sqrt{3}}s + \frac{1}{\sqrt{6}}\{p_x(\varepsilon + \varepsilon^*) + ip_y(\varepsilon - \varepsilon^*)\}$$

$$= \frac{1}{\sqrt{3}}s - \frac{1}{\sqrt{6}}p_x - \frac{1}{\sqrt{2}}p_y$$

$$\phi_3 = \frac{1}{\sqrt{3}}\left\{s + \frac{\varepsilon}{\sqrt{2}}(p_x - ip_y) + \frac{\varepsilon^*}{\sqrt{2}}(p_x + ip_y)\right\}$$

$$= \frac{1}{\sqrt{3}}s - \frac{1}{\sqrt{6}}p_x + \frac{1}{\sqrt{2}}p_y$$

We could have employed C_{3v} just as easily.

For tetrahedral CH_4 we play a trick similar to what was done for linear CH_2. The simplest group that separates the p orbitals is again D_2, and it is a subgroup of the cubic groups. To visualize the hybrid directions more easily we show them as lines pointing toward the corners of a cube (see Figure 9 and Table 4).

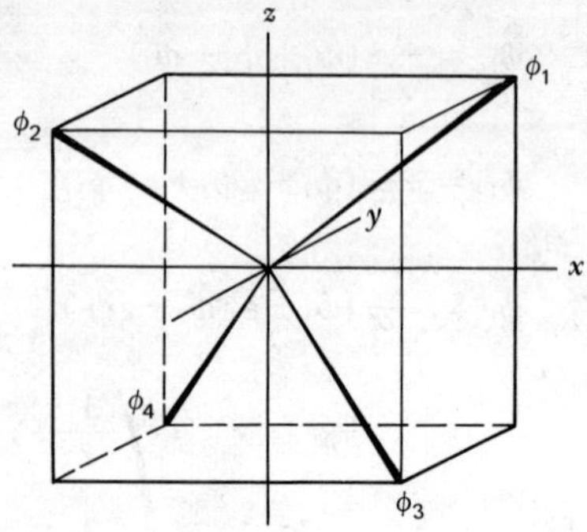

Figure 9

Table 4

D_2	E	$C_2^{(z)}$	$C_2^{(y)}$	$C_2^{(x)}$	
A	1	1	1	1	s
B_1	1	1	-1	-1	z
B_2	1	-1	1	-1	y
B_3	1	-1	-1	1	x
ϕ_1	ϕ_1	ϕ_2	ϕ_4	ϕ_3	
ϕ_2	ϕ_2	ϕ_1	ϕ_3	ϕ_4	
ϕ_3	ϕ_3	ϕ_4	ϕ_2	ϕ_1	
ϕ_4	ϕ_4	ϕ_3	ϕ_1	ϕ_1	
Γ	4	0	0	0	

$$\Gamma = A + B_1 + B_2 + B_3$$

$$\psi_A = \tfrac{1}{2}(\phi_1 + \phi_2 + \phi_3 + \phi_4)$$
$$\psi_{B_1} = \tfrac{1}{2}(\phi_1 + \phi_2 - \phi_3 - \phi_4)$$
$$\psi_{B_2} = \tfrac{1}{2}(\phi_1 - \phi_2 + \phi_3 - \phi_4)$$
$$\psi_{B_3} = \tfrac{1}{2}(\phi_1 - \phi_2 - \phi_3 + \phi_4)$$

The transformation matrix is easily written down

$$(\psi_A, \psi_{B_1}, \psi_{B_2}, \psi_{B_3}) = (\phi_1, \phi_2, \phi_3, \phi_4)\begin{pmatrix} \frac{1}{2} & \frac{1}{2} & \frac{1}{2} & \frac{1}{2} \\ \frac{1}{2} & \frac{1}{2} & -\frac{1}{2} & -\frac{1}{2} \\ \frac{1}{2} & -\frac{1}{2} & \frac{1}{2} & -\frac{1}{2} \\ \frac{1}{2} & -\frac{1}{2} & -\frac{1}{2} & +\frac{1}{2} \end{pmatrix}$$

and we find for the tetrahedral hybrids

$$\phi_1 = \tfrac{1}{2}(s + p_z + p_y + p_x)$$
$$\phi_2 = \tfrac{1}{2}(s + p_z - p_y - p_x)$$
$$\phi_3 = \tfrac{1}{2}(s - p_z + p_y - p_x)$$
$$\phi_4 = \tfrac{1}{2}(s - p_z - p_y + p_x)$$

Appendix A
Some Aspects of Class Algebra

In this appendix we demonstrate the column orthogonality of the character table of irreducible representations. Also, some interesting properties of classes, class algebra, and a means of directly determining characters for irreducible representations will be demonstrated.

Let $s_j^{(i)}$ be one of the elements of G belonging to K_i, the ith class (a set). To find K_i we pick some $s_j^{(i)}$, form the products $g^{-1}s_j^{(i)}g$ for all $g \in G$ and collect the distinct elements resulting from the conjugation process. Having worked through several examples the reader will notice that in performing such a process each $s_j^{(i)} \in K_i$ appears as often as every other element in the ith class. This is so because, as we show below,

$$\sum_{g \in G} g^{-1}s_j^{(i)}g = \frac{^oG}{^oK_i} K_i$$

$$K_i = \sum_j s_j^{(i)} \tag{A.1}$$

where ^oK_i is the order of the ith class. To demonstrate Eq. (A.1) we show first that

$$\sigma_j = \sum_{g \in G} g^{-1}s_j^{(i)}g = \sigma \tag{A.2}$$

is indeed independent of j. Let $s_j^{(i)}$ and $s_q^{(i)}$ be conjugate by the element r. Then

$$\sigma_j = \sum_{g \in G} g^{-1}s_j^{(i)}g$$

$$= \sum_{g \in G} g^{-1}r^{-1}s_q^{(i)}rg$$

$$= \sum_{y \in G} y^{-1}s_q^{(i)}y = \sigma_q \tag{A.3}$$

where $y = rg$ and the sum over g of rg is equivalent to the sum over the product (rg) by the rearrangement theorem. Thus $\sigma_j = \sigma_q = \sigma$ and is independent of the choice of j (or q). That Eq. (A.1) is true now follows from the invariance property of classes with respect to conjugation. Thus, for any $t \in G$

$$t^{-1}K_it = \sum_j t^{-1}s_j^{(i)}t = K_i \tag{A.4}$$

since all the products $t^{-1}s_j^{(i)}t$ must be distinct (otherwise $t^{-1}s_j^{(i)}t = t^{-1}s_k^{(i)}t$ implies $s_j^{(i)} = s_k^{(i)}$, violating the uniqueness property of the group elements) and must, of course, belong to K_i. It is clear then that

$$\sum_{t \in G} t^{-1}K_it = \sum_{t \in G} K_i = (°G)K_i \tag{A.5}$$

Also,

$$\begin{aligned}\sum_{t \in G} t^{-1}K_it &= \sum_j \left(\sum_{t \in G} t^{-1}s_j^{(i)}t\right) \\ &= \sum_j \sigma_j = \sigma \sum_j 1 \\ &= (°K_i)\left(\sum_{t \in G} t^{-1}s_j^{(i)}t\right)\end{aligned} \tag{A.6}$$

where the final step follows from the fact that $\sigma = \sum_{t \in G} t^{-1}s_j^{(i)}t$ is, as shown earlier, independent of j. Thus, equating (A.5) and (A.6) we arrive at the result given in Eq. (A.1).

Note in addition that since σ contains all the elements of K_i the same number of times and since this number must be an integer, we have also shown that $°G \div °K_i$ is integral; that is, the order of a class is a divisor of the order of the group.

Consider next the product of classes

$$K_iK_j = \sum_{n,m} s_n^{(i)} s_m^{(j)} \tag{A.7}$$

Since, for any $t \in G$, $t^{-1}K_it = K_i$, it follows that

$$t^{-1}K_iK_jt = t^{-1}K_itt^{-1}K_jt = K_iK_j \tag{A.8}$$

Thus, the product K_iK_j is invariant under conjugation and must therefore consist of complete classes

$$K_iK_j = \sum_k C_{ij}^k K_k \tag{A.9}$$

where the coefficients C_{ij}^k are called the *class constants*.

Since a representation is a homomorphic mapping of the group it is necessary for the matrices representing G to also obey the equations derived above. Thus, let

$$\mathbf{D}_i = \sum_j \mathbf{D}(s_j^{(i)}) \tag{A.10}$$

be the matrix sum of the matrices of elements in the ith class. Then, by (A.9)

$$\mathbf{D}_i\mathbf{D}_j = \sum_k C_{ij}^k \mathbf{D}_k \tag{A.11}$$

Also, from (A.4)

$$\mathbf{D}(t)\mathbf{D}_i = \mathbf{D}_i\mathbf{D}(t), \qquad \text{all} \qquad t \in G \tag{A.12}$$

where $\mathbf{D}(t)$ is the matrix representative of the element t. Equation (A.12) is quite significant for if we consider the αth *irreducible* representation of G it says that $\mathbf{D}_i^{(\alpha)}$—the summed class matrix in Γ_α—commutes with all matrices $\mathbf{D}^{(\alpha)}(t)$ of the irreducible representation. By Schur's first lemma this means $\mathbf{D}_i^{(\alpha)}$ must be a multiple of the unit matrix

$$\mathbf{D}_i^{(\alpha)} = \lambda_i^{(\alpha)}\mathbf{I} \tag{A.13}$$

The constant $\lambda_i^{(\alpha)}$ can be determined by taking the trace of $\mathbf{D}_i^{(\alpha)}$ and using the fact that the character of all elements in the same class is the same.

$$\begin{aligned} \operatorname{Tr} \mathbf{D}_i^{(\alpha)} &= ({}^o\mathrm{K}_i)\chi_i^{(\alpha)} = \operatorname{Tr} \lambda_i^{(\alpha)}\mathbf{I} \\ &= \lambda_i^{(\alpha)} d_\alpha \\ \lambda_i^{(\alpha)} &= \chi_i^{(\alpha)} \frac{{}^o\mathrm{K}_i}{d_\alpha} \end{aligned} \tag{A.14}$$

where d_α is the dimension of the αth irreducible representation. For the class matrices for irreducible representations (A.11) becomes

$$\begin{aligned} \frac{{}^o\mathrm{K}_i}{d_\alpha}\chi_i^{(\alpha)} \frac{{}^o\mathrm{K}_j}{d_\alpha}\chi_j^{(\alpha)}\mathbf{II} &= \sum_k C_{ij}^k \frac{{}^o\mathrm{K}_k}{d_\alpha}\chi_k^{(\alpha)}\mathbf{I} \\ ({}^o\mathrm{K}_i {}^o\mathrm{K}_j)\chi_i^{(\alpha)}\chi_j^{(\alpha)} &= d_\alpha \sum_k C_{ij}^k ({}^o\mathrm{K}_k)\chi_k^{(\alpha)} \end{aligned} \tag{A.15}$$

and we now have an equation relating the products of characters of the irreducible representations. The class constants can be worked out from the properties of the abstract group, and d_α and ${}^o\mathrm{K}_i$ are readily determined; application of the various equations embodied in (A.15) (for various i and j) then leads to a set of equations from which the individual characters can be derived. It is sometimes a tedious process, but it is a direct one that will yield unambiguous results.

It is worthwhile to illustrate the process briefly here using our old friend $G = C_{3v}$. Let $K_1 = \{E\}$, $K_2 = \{C_2, C_2^2\}$, $K_3 = \{\sigma_v^{(1)}, \sigma_v^{(2)}, \sigma_v^{(3)}\}$. Then

$$K_1K_i = K_i \qquad \text{(nothing new!)}$$

(1) $K_2^2 = 2K_1 + K_2$

(2) $K_2K_3 = 2K_3$

(3) $K_3K_3 = 3K_1 + 3K_2$

The d_α are 1, 1, and 2; for general $d_\alpha \equiv d = \chi_1^{(\alpha)}$ we have then

$$(1) \qquad 4\chi_2^2 = d\{2 \cdot 1 \cdot d + 1 \cdot 2 \cdot \chi_2\}$$

$$\chi_2 = d, -d/2$$

$$(2) \quad 2 \cdot 3 \cdot \chi_2\chi_3 = d \cdot 2 \cdot 3 \cdot \chi_3$$

$$\chi_2 = d \text{ or } \chi_3 = 0$$

$$(3) \qquad 3 \cdot 3\chi_3^2 = d\{3 \cdot 1 \cdot d + 3 \cdot 2 \cdot \chi_2\}$$

For $d = 1$ the characters must be of modulus 1 ($\chi_2 \neq \frac{1}{2}$, $\chi_3 \neq 0$) so that

$$\chi_1 = 1, \qquad \chi_2 = 1, \qquad \chi_3 = \pm 1 \qquad (d = 1)$$

For $d = 2$, if $\chi_2 = d = 2$ from (1) then (3) leads to $\chi_3 = 2$. But this result violates the "normalization requirement" for irreducible representations

$$\langle \chi^{(\alpha)} | \chi^{(\alpha)} \rangle = \frac{1}{{}^oG} \sum_i {}^oK_i \chi_i^{(\alpha)*} \chi_i^{(\alpha)} = 1$$

Thus, $\chi_2 = -d/2 = -1$ and $\chi_3 = 0$ must be the correct solution

$$\chi_1 = 2, \qquad \chi_2 = -1, \qquad \chi_3 = 0 \qquad (d = 2)$$

We have found the character for the three irreducible representations. Although the quadratic nature of the class constants equation may present an indeterminacy in choosing the $\chi_i^{(\alpha)}$, this is removed by verifying the character normalization condition.

Our final point in this discussion is to demonstrate the orthogonality of the columns of character tables of irreducible representations. Equation (A.15) has the type of product on the left we want to manipulate. Indeed, we want such products summed over all the irreducible representations:

$$\begin{aligned} \sum_\alpha ({}^oK_i)\chi_i^{(\alpha)}({}^oK_j)\chi_j^{(\alpha)} &= d_\alpha \sum_{k,\alpha} C_{ij}^k({}^oK_k)\chi_k^{(\alpha)} \\ &= \sum_k C_{ij}^k({}^oK_k) \sum_\alpha \chi^{(\alpha)}(e)\chi_k^{(\alpha)} \end{aligned} \tag{A.16}$$

where we have used $d_\alpha = \chi_1^{(\alpha)} = \chi^{(\alpha)}(e)$ on the right-hand side of (A.16). The particular form on the right is desirable since, from the regular representation, we know

$$\begin{aligned} \chi^R(s) &= 0, \qquad s \neq e \\ \chi^R(s) &= \sum_\alpha c_\alpha^R \chi^{(\alpha)}(s) \\ c_\alpha^R &= d_\alpha = \chi^{(\alpha)}(e) \\ \sum_\alpha \chi^{(\alpha)}(e)\chi^{(\alpha)}(s) &= \delta_{s,e}{}^oG \qquad (\text{since } \chi^R(e) = {}^oG) \end{aligned} \tag{A.17}$$

Equation (A.16) then becomes ($K_1 = \{e\}$, ${}^oK_1 = 1$ always)

$$\sum_\alpha ({}^oK_i)\chi_i^{(\alpha)}({}^oK_j)\chi_j^{(\alpha)} = C_{ij}^1 \cdot 1 \cdot {}^oG \tag{A.18}$$

Now the only product K_iK_j that *can* contain K_1 will be those where K_j is the class of the inverses of the elements of K_i. It is readily seen that all such inverses must belong to the same class; furthermore, in the products $a_n^{(i)}a_m^{(j)}$, e will occur *only* for $a_n^{(i)} = [a_m^{(j)}]^{-1}$ and will occur exactly ${}^oK_1$ times: $C_{ij}^1 = {}^oK_i$ for K_j = class of inverses of elements of K_i. Finally, recalling that $\chi(s^{-1}) = \chi(s)^{-1} = \chi(s)^*$, (A.18) becomes

$$\begin{aligned} \sum_\alpha ({}^oK_i)\chi_i^{(\alpha)}({}^oK_j)\chi_j^{(\alpha)*} &= {}^oK_i{}^oG\delta_{ij} \\ \sum_\alpha \chi_i^{(\alpha)}\chi_j^{(\alpha)*} &= \frac{{}^oG}{{}^oK_i}\,\delta_{ij} \end{aligned} \tag{A.19}$$

which demonstrates the column orthogonality of the character table. In the form

$$\sum_\alpha \left(\frac{{}^oK_i}{{}^oG}\right)^{1/2} \chi_i^{(\alpha)} \cdot \left(\frac{{}^oK_j}{{}^oG}\right)^{1/2} \chi_j^{(\alpha)*} = \delta_{ij} \tag{A.20}$$

we see that the quantities $({}^oK_i/{}^oG)^{1/2}\chi_i^{(\alpha)}$ act as the components of normalized and orthogonal class vectors in a space whose dimension is equal to the number of irreducible representations of the group. By the properties of vector spaces we can have no more linearly independent orthogonal vectors than the dimension of the space in question so we know that the number of classes must be *less than or equal to* the number of irreducible representations. In Chapter 9 just the opposite inequality was shown; taken together, then, the equality must obtain so that *the number of classes equals the number of irreducible representations.*

Appendix B
Symmetry Reducing Perturbations and Group Correlation Tables

Another example of group theory applied to quantum mechanics is the important and very useful topic of the effects of symmetry reducing perturbations. We want to know what happens to a system of states characterized by some given symmetry when a perturbation is introduced that lowers the overall symmetry of the system. We have actually looked at this very briefly once before in problem 12.10 when we examined the predicted degeneracies (predicted by the irreducibility postulate) of the $3d$ orbitals in the groups D_2, D_4, and D_{4h} of increasing symmetry. The problem is stated very simply: What happens to the symmetry classification of a state when the symmetry of a system is lowered and what are the physical consequences?

Of course, when the symmetry is lowered (or raised) the notation may change but this is a matter of bookkeeping rather than the physics (or chemistry) of the situation. More important is the question of the removal of degeneracy. Thus, a level or state that is characterized by E symmetry in D_4 must have its twofold degeneracy removed when the symmetry is reduced to D_2 simply because there are no two-dimensional representations among the irreducible representations of D_2 but only one-dimensional ones. Since we are speaking of the effect of perturbations we shall restrict ourselves to those effects which reduce the group G to one of its subgroups $H \subset G$. This makes physical sense; a small shift of atomic positions can remove a C_4-axis (perhaps changing it into a C_2-axis as in the above example) but it would hardly be a small effect to convert a C_4-axis into a C_3-axis.

The limitation that the lower symmetry group H is a subgroup of G is the key to determining the change in symmetry classification of the state in question. Thus, since the state transforms as $\Gamma_\alpha^{(G)}$ in G with characters $\chi^{(\alpha;G)}(s)$, $s \in G$ and since $H \subset G$ then $\Gamma_\alpha^{(G)}$ must also be a representation of H with characters $\chi^{(\alpha;G)}(s)$, $s \in H$. That this is so simply follows from the fact that $\mathbf{D}^{(\alpha;G)}(s)\mathbf{D}^{(\alpha;G)}(t) = \mathbf{D}^{(\alpha;G)}(st)$ for all $s, t \in G$ and so must hold for all $h \in H \subset G$. The representation for H so obtained may or may not be an irreducible representation, and, indeed, it is on this that the degeneracy question depends. To find the decomposition of $\Gamma_\alpha^{(G)}$ in H we proceed as usual using the $\chi^{(\alpha;G)}(s)$, $s \in H$ and the character orthogonality conditions appropriate to H.

Consider the above-mentioned example of changing D_4 to D_2 when we want to find what happens to the Γ_E representation in D_{4h}. Removal of the C_4-axis (which also removes two of the four perpendicular C_2-axes) by reducing it to only a C_2-axis yields a representation of $H = D_2$, where the characters are simply those of the remaining elements: $\chi(E)=2$, $\chi(C_2)=-2$, $\chi(C_2')=\chi(C_2'')=0$. These elements correspond to E, $C_2^{(z)}$, $C_2^{(y)}$, and $C_2^{(x)}$ of D_2, and it is clear that

$$E \text{ in } D_4 \rightarrow B_2 + B_3 \text{ in } D_2$$

We knew beforehand, of course, that the degeneracy had to be removed and now we have found the resulting new states.

This type of manipulation, by which irreducible representations of a group are correlated with those of its subgroups, leads to what are termed *group correlation tables*. These are simply tabulations of the character decompositions as were performed above. Sometimes more than one correlation is possible if we have a choice of equivalent symmetry elements to retain; in such cases we must make the necessary specifications. Suppose we wish to construct the group correlation table for D_3, a particularly simple example.

D_3	E	$2C_3$	$3C_2$
A_1	1	1	1
A_2	1	1	−1
E	2	−1	0

C_3		E	C_3	C_3^2
A		1	1	1
E	{	1	ε	ε^*
		1	ε^*	ε
		$\varepsilon = e^{2\pi i/3}$		

C_2	E	C_2
A	1	1
B	1	−1

D_3 has the nontrivial subgroup C_3 and C_2, and since the twofold axes are all in the same class there will be a unique $D_3 \rightarrow C_2$ correlation. For the subgroup C_3 we use just the characters for E and $2C_3$ in D_3 to find that $A_1 \rightarrow A$, $A_2 \rightarrow A$, and $E \rightarrow E$. For the subgroup C_2 we employ only the characters of E and $3C_2$ (only *one* C_2 character) to obtain $A_1 \rightarrow A$, $A_2 \rightarrow B$,

$E \to A + B$. We would tabulate the above correlation as

D_3	C_3	C_2
A_1	A	A
A_2	A	B
E	E	$A+B$

where the first row labels the group and its subgroups. Extensive group correlation tables may be found, for example, in E. B. Wilson, J. C. Decius, and P. C. Cross, *Molecular Vibrations*, McGraw-Hill, New York, 1955, in G. F. Koster, J. O. Dimmock, R. G. Wheeler, and H. Statz, *Properties of the Thirty-two Point Groups*, The M.I.T. Press, Cambridge, 1963, or in the handy pamphlet by P.W. Atkins, M. S. Child, and C. S. G. Phillips, *Tables for Group Theory*, Oxford University Press, Oxford, 1970. The following examples provide some additional experience with manipulations of this sort.

What happens to the degeneracy of E and T_1 states in the group O when the symmetry is reduced to D_4? *We must list the characters of O* for the states in question that remain under the reduced symmetry.

D_4		E	$2C_4$	$C_2(=C_4^2)$	$2C_2'$	$2C_2''$
O	E	2	0	2	2	0
	T_1	3	1	−1	−1	−1

$E \to A_1 + B_1$ and $T_1 \to A_2 + E$ so that the degeneracy of the E state is completely removed while the threefold degeneracy of the T_1 state is only partly lifted. The trick here is to recognize that the $3C_2(=C_4^2)$ operations in O go with $C_2(=C_4^2)$ *and* $2C_2'$ in D_4 while the character for the $6C_2$ set in O goes over into that for $2C_2''$ in D_4. It is instructive in this example to sketch the symmetry elements in the group O and see exactly which remain in D_4.

Finally, let us construct the correlation tables for C_4 and C_{3v}. C_4 has but one nontrivial subgroup C_2. Thus, using the same type of table as was employed in the previous example we have

C_2		E	C_2
C_4	A	1	1
	B	1	1
	E	1	−1
		1	−1

C_4	C_2
A	A
B	A
E	$2B$

and the resulting correlation tables are clear.

For C_{3v} the subgroups to be considered are C_3 and C_s. Thus

C_3		E	C_3	C_3^2
C_{3v}	A_1	1	1	1
	A_2	1	1	1
	E	2	−1	−1

C_s		E	σ
C_{3v}	A_1	1	1
	A_2	1	−1
	E	2	0

with the correlation table

C_{3v}	C_3	C_s
A_1	A	A'
A_2	A	A''
E	E	$A' + A''$

Appendix C
Selected Character Tables

This appendix contains the character tables for the 32 crystallographic point groups (those point groups allowable in a crystal lattice) plus those for the groups C_5, C_7, C_8, D_5, C_{5v}, C_{5h}, D_{5h}, D_{4d}, D_{5d}, D_{6d}, S_8, I, I_h, $C_{\infty v}$, and $D_{\infty h}$ which are also often important in symmetry considerations of molecules. Schönflies notation for the point groups is employed and the reader is referred to Chapter 9 for a discussion of the notation for the irreducible representations. Pairs of complex-conjugate representations are bracketed together and labeled as a two-dimensional representation E as discussed in Chapters 9 and 12.

1. The Nonaxial Groups C_1, C_s, C_i

C_1	E
A	1

$C_s = C_h$	E	σ_h		
A'	1	1	x, y, R_z	x^2, y^2, z^2, xy
A''	1	−1	z, R_x, R_y	yz, xz

$C_i = S_2$	E	i		
A_g	1	1	R_x, R_y, R_z	$x^2, y^2, z^2, xy, xz, yz$
A_u	1	−1	x, y, z	

2. The C_n Groups ($n = 2, 3, \ldots, 8$)

C_2	E	C_2		
A	1	1	z, R_z	x^2, y^2, z^2, xy
B	1	−1	x, y, R_x, R_y	yz, xz

C_3	E	C_3	C_3^2	$\varepsilon = \exp(2\pi i/3)$	
A	1	1	1	z, R_z	x^2+y^2, z^2
E	$\begin{cases}1\\1\end{cases}$	ε ε^*	ε^* ε	$(x, y)(R_x, R_y)$	$(x^2-y^2, xy)(yz, xz)$

C_4	E	C_4	C_2	C_4^3		
A	1	1	1	1	z, R_z	x^2+y^2, z^2
B	1	−1	1	−1		x^2-y^2, xy
E	$\begin{cases}1\\1\end{cases}$	i $-i$	−1 −1	$-i$ i	$(x, y)(R_x, R_y)$	(yz, xz)

C_5	E	C_5	C_5^2	C_5^3	C_5^4	$\varepsilon = \exp(2\pi i/5)$	
A	1	1	1	1	1	z, R_z	x^2+y^2, z^2
E_1	$\begin{cases}1\\1\end{cases}$	ε ε^*	ε^2 ε^{*2}	ε^{*2} ε^2	ε^* ε	$(x, y)(R_x, R_y)$	(yz, xz)
E_2	$\begin{cases}1\\1\end{cases}$	ε^2 ε^{*2}	ε^* ε	ε ε^*	ε^{*2} ε^2		(x^2-y^2, xy)

C_6	E	C_6	C_3	C_2	C_3^2	C_6^5	$\varepsilon = \exp(2\pi i/6)$	
A	1	1	1	1	1	1	z, R_z	x^2+y^2, z^2
B	1	−1	1	−1	1	−1		
E_1	$\begin{cases}1\\1\end{cases}$	ε ε^*	$-\varepsilon^*$ $-\varepsilon$	−1 −1	$-\varepsilon$ $-\varepsilon^*$	ε^* ε	(x, y) (R_z, R_y)	(xz, yz)
E_2	$\begin{cases}1\\1\end{cases}$	$-\varepsilon^*$ $-\varepsilon$	$-\varepsilon$ $-\varepsilon^*$	1 1	$-\varepsilon^*$ $-\varepsilon^*$	$-\varepsilon$ $-\varepsilon^*$		(x^2-y^2, xy)

C_7	E	C_7	C_7^2	C_7^3	C_7^4	C_7^5	C_7^6	$\varepsilon = \exp(2\pi i/7)$	
A	1	1	1	1	1	1	1	z, R_z	x^2+y^2, z^2
E_1	$\begin{cases}1\\1\end{cases}$	ε ε^*	ε^2 ε^{*2}	ε^3 ε^{*3}	ε^{*3} ε^3	ε^{*2} ε^2	ε^* ε	(x, y) (R_x, R_y)	(xz, yz)
E_2	$\begin{cases}1\\1\end{cases}$	ε^2 ε^{*2}	ε^{*3} ε^3	ε^* ε	ε ε^*	ε^3 ε^{*3}	ε^{*2} ε^2		(x^2-y^2, xy)
E_3	$\begin{cases}1\\1\end{cases}$	ε^3 ε^{*3}	ε^* ε	ε^2 ε^{*2}	ε^{*2} ε^2	ε ε^*	ε^{*3} ε^3		

C_8	E	C_8	C_4	C_2	C_4^3	C_8^3	C_8^5	C_8^7	$\varepsilon = \exp(2\pi i/8)$	
A	1	1	1	1	1	1	1	1	z, R_z	x^2+y^2, z^2
B	1	−1	1	1	1	−1	−1	−1		
E_1	$\begin{Bmatrix}1\\1\end{Bmatrix}$	$\begin{matrix}\varepsilon\\\varepsilon^*\end{matrix}$	$\begin{matrix}i\\-i\end{matrix}$	$\begin{matrix}-1\\-1\end{matrix}$	$\begin{matrix}-i\\i\end{matrix}$	$\begin{matrix}-\varepsilon^*\\-\varepsilon\end{matrix}$	$\begin{matrix}-\varepsilon\\-\varepsilon^*\end{matrix}$	$\begin{matrix}\varepsilon^*\\\varepsilon\end{matrix}$	(x, y) (R_x, R_y)	(xz, yz)
E_2	$\begin{Bmatrix}1\\1\end{Bmatrix}$	$\begin{matrix}i\\-i\end{matrix}$	$\begin{matrix}-1\\-1\end{matrix}$	$\begin{matrix}1\\1\end{matrix}$	$\begin{matrix}-1\\-1\end{matrix}$	$\begin{matrix}-i\\i\end{matrix}$	$\begin{matrix}i\\-i\end{matrix}$	$\begin{matrix}-i\\i\end{matrix}$		(x^2-y^2, xy)
E_3	$\begin{Bmatrix}1\\1\end{Bmatrix}$	$\begin{matrix}-\varepsilon\\-\varepsilon^*\end{matrix}$	$\begin{matrix}i\\-i\end{matrix}$	$\begin{matrix}-1\\-1\end{matrix}$	$\begin{matrix}-i\\i\end{matrix}$	$\begin{matrix}\varepsilon^*\\\varepsilon\end{matrix}$	$\begin{matrix}\varepsilon\\\varepsilon^*\end{matrix}$	$\begin{matrix}-\varepsilon^*\\-\varepsilon\end{matrix}$		

3. The D_n Groups ($n = 2, 3, 4, 5, 6$)

D_2	E	$C_2(z)$	$C_2(y)$	$C_2(x)$		
A	1	1	1	1		x^2, y^2, z^2
B_1	1	1	−1	−1	z, R_z	xy
B_2	1	−1	1	−1	y, R_y	xz
B_3	1	−1	−1	1	x, R_x	yz

D_3	E	$2C_3$	$3C_2$		
A_1	1	1	1		x^2+y^2, z^2
A_2	1	1	−1	z, R_z	
E	2	−1	0	$(x, y)(R_x, R_y)$	$(x^2-y^2, xy)(xz, yz)$

D_4	E	$2C_4$	$C_2(=C_4^2)$	$2C_2'$	$2C_2''$		
A_1	1	1	1	1	1		x^2+y^2, z^2
A_2	1	1	1	−1	−1	z, R_z	
B_1	1	−1	1	1	−1		x^2-y^2
B_2	1	−1	1	−1	1		xy
E	2	0	−2	0	0	$(x, y)(R_x, R_y)$	(xz, yz)

D_5	E	$2C_5$	$2C_5^2$	$5C_2$		
A_1	1	1	1	1		x^2+y^2, z^2
A_2	1	1	1	−1	z, R_z	
E_1	2	2 cos 72°	2 cos 144°	0	$(x, y)(R_x, R_y)$	(xz, yz)
E_2	2	2 cos 144°	2 cos 72°	0		(x^2-y^2, xy)

D_6	E	$2C_6$	$2C_3$	C_2	$3C_2'$	$3C_2''$		
A_1	1	1	1	1	1	1		x^2+y^2, z^2
A_2	1	1	1	1	−1	−1	z, R_z	
B_1	1	−1	1	−1	1	−1		
B_2	1	−1	1	−1	−1	1		
E_1	2	1	−1	−2	0	0	$(x, y)(R_x, R_y)$	(xz, yz)
E_2	2	−1	−1	2	0	0		(x^2-y^2, xy)

4. The C_{nv} Groups ($n = 2, 3, 4, 5, 6$)

C_{2v}	E	C_2	$\sigma_v(xz)$	$\sigma_v'(yz)$		
A_1	1	1	1	1	z	x^2, y^2, z^2
A_2	1	1	−1	−1	R_z	xy
B_1	1	−1	1	−1	x, R_y	xz
B_2	1	−1	−1	1	y, R_x	yz

C_{3v}	E	$2C_3$	$3\sigma_v$		
A_1	1	1	1	z	x^2+y^2, z^2
A_2	1	1	−1	R_z	
E	2	−1	0	$(x, y)(R_x, R_y)$	$(x^2-y^2, xy)(xz, yz)$

C_{4v}	E	$2C_4$	C_2	$2\sigma_v$	$2\sigma_d$		
A_1	1	1	1	1	1	z	x^2+y^2, z^2
A_2	1	1	1	−1	−1	R_z	
B_1	1	−1	1	1	−1		x^2-y^2
B_2	1	−1	1	−1	1		xy
E	2	0	−2	0	0	$(x, y)(R_x, R_y)$	(xz, yz)

C_{5v}	E	$2C_5$	$2C_5^2$	$5\sigma_v$		
A_1	1	1	1	1	z	x^2+y^2, z^2
A_2	1	1	1	−1	R_z	
E_1	2	2 cos 72°	2 cos 144°	0	$(x, y)(R_x, R_y)$	(xz, yz)
E_2	2	2 cos 144°	2 cos 72°	0		(x^2-y^2, xy)

C_{6v}	E	$2C_6$	$2C_3$	C_2	$3\sigma_v$	$3\sigma_d$		
A_1	1	1	1	1	1	1	z	x^2+y^2, z^2
A_2	1	1	1	1	−1	−1	R_z	
B_1	1	−1	1	−1	1	−1		
B_2	1	−1	1	−1	−1	1		
E_1	2	1	−1	−2	0	0	$(x, y)(R_x, R_y)$	(xz, yz)
E_2	2	−1	−1	2	0	0		(x^2-y^2, xy)

5. The C_{nh} Groups ($n = 2, 3, 4, 5, 6$)

C_{2h}	E	C_2	i	σ_h		
A_g	1	1	1	1	R_z	x^2, y^2, z^2, xy
B_g	1	−1	1	−1	R_x, R_y	xz, yz
A_u	1	1	−1	−1	z	
B_u	1	−1	−1	1	x, y	

C_{3h}	E	C_3	C_3^2	σ_h	S_3	S_3^5	$\varepsilon = \exp(2\pi i/3)$	
A'	1	1	1	1	1	1	R_z	x^2+y^2, z^2
E'	1	ε	ε^*	1	ε	ε^*	(x, y)	(x^2-y^2, xy)
	1	ε^*	ε	1	ε^*	ε		
A''	1	1	1	−1	−1	−1	z	
E''	1	ε	ε^*	−1	$-\varepsilon$	$-\varepsilon^*$	(R_x, R_y)	(xz, yz)
	1	ε^*	ε	−1	$-\varepsilon^*$	$-\varepsilon$		

C_{4h}	E	C_4	C_2	C_4^3	i	S_4^3	σ_h	S_4		
A_g	1	1	1	1	1	1	1	1	R_z	x^2+y^2, z^2
B_g	1	−1	1	−1	1	−1	1	−1		x^2-y^2, xy
E_g	1	i	−1	$-i$	1	i	−1	$-i$	(R_x, R_y)	(xz, yz)
	1	$-i$	−1	i	1	$-i$	−1	i		
A_u	1	1	1	1	−1	−1	−1	−1	z	
B_u	1	−1	1	−1	−1	1	−1	1		
E_u	1	i	−1	$-i$	−1	$-i$	1	i	(x, y)	
	1	$-i$	−1	i	−1	i	1	$-i$		

C_{5h}	E	C_5	C_5^2	C_5^3	C_5^4	σ_h	S_5	S_5^7	S_5^3	S_5^9	$\varepsilon = \exp(2\pi i/5)$	
A'	1	1	1	1	1	1	1	1	1	1	R_z	x^2+y^2, z^2
E_1'	1	ε	ε^2	ε^{*2}	ε^*	1	ε	ε^2	ε^{*2}	ε^*	(x, y)	
	1	ε^*	ε^{*2}	ε^2	ε	1	ε^*	ε^{*2}	ε^2	ε		
E_2'	1	ε^2	ε^*	ε	ε^{*2}	1	ε^2	ε^*	ε	ε^{*2}		(x^2-y^2, xy)
	1	ε^{*2}	ε	ε^*	ε^2	1	ε^{*2}	ε	ε^*	ε^2		
A''	1	1	1	1	1	−1	−1	−1	−1	−1	z	
E_1''	1	ε	ε^2	ε^{*2}	ε^*	−1	$-\varepsilon$	$-\varepsilon^2$	$-\varepsilon^{*2}$	$-\varepsilon^*$	(R_x, R_y)	(xz, yz)
	1	ε^*	ε^{*2}	ε^2	ε	−1	$-\varepsilon^*$	$-\varepsilon^{*2}$	$-\varepsilon^2$	$-\varepsilon$		
E_2''	1	ε^2	ε^*	ε	ε^{*2}	−1	$-\varepsilon^2$	$-\varepsilon^*$	$-\varepsilon$	$-\varepsilon^{*2}$		
	1	ε^{*2}	ε	ε^*	ε^2	−1	$-\varepsilon^{*2}$	$-\varepsilon$	$-\varepsilon^*$	$-\varepsilon^2$		

C_{6h}	E	C_6	C_3	C_2	C_3^2	C_6^5	i	S_3^5	S_6^5	σ_h	S_6	S_3	$\varepsilon = \exp(2\pi i/6)$	
A_g	1	1	1	1	1	1	1	1	1	1	1	1	R_z	x^2+y^2, z^2
B_g	1	−1	1	−1	1	−1	1	−1	1	−1	1	−1		
E_{1g}	$\{1$	ε	$-\varepsilon^*$	−1	$-\varepsilon$	ε^*	1	ε	$-\varepsilon^*$	−1	$-\varepsilon$	$\varepsilon^*\}$	(R_x, R_y)	(xz, yz)
	$\{1$	ε^*	$-\varepsilon$	−1	$-\varepsilon^*$	ε	1	ε^*	$-\varepsilon$	−1	$-\varepsilon^*$	$\varepsilon\}$		
E_{2g}	$\{1$	$-\varepsilon^*$	$-\varepsilon$	1	$-\varepsilon^*$	$-\varepsilon$	1	$-\varepsilon^*$	$-\varepsilon$	1	$-\varepsilon^*$	$-\varepsilon\}$		(x^2-y^2, xy)
	$\{1$	$-\varepsilon$	$-\varepsilon^*$	1	$-\varepsilon$	$-\varepsilon^*$	1	$-\varepsilon$	$-\varepsilon^*$	1	$-\varepsilon$	$-\varepsilon^*\}$		
A_u	1	1	1	1	1	1	−1	−1	−1	−1	−1	−1	z	
B_u	1	−1	1	−1	1	−1	−1	1	−1	1	−1	1		
E_{1u}	$\{1$	ε	$-\varepsilon^*$	−1	$-\varepsilon$	ε^*	−1	$-\varepsilon$	ε^*	1	ε	$-\varepsilon^*\}$	(x, y)	
	$\{1$	ε^*	$-\varepsilon$	−1	$-\varepsilon^*$	ε	−1	$-\varepsilon^*$	ε	1	ε^*	$-\varepsilon\}$		
E_{2u}	$\{1$	$-\varepsilon^*$	$-\varepsilon$	1	$-\varepsilon^*$	$-\varepsilon$	−1	ε^*	ε	−1	ε^*	$\varepsilon\}$		
	$\{1$	$-\varepsilon$	$-\varepsilon^*$	1	$-\varepsilon$	$-\varepsilon^*$	−1	ε	ε^*	−1	ε	$\varepsilon^*\}$		

6. The D_{nh} Groups ($n = 2, 3, 4, 5, 6$)

D_{2h}	E	$C_2(z)$	$C_2(y)$	$C_2(x)$	i	$\sigma(xy)$	$\sigma(xz)$	$\sigma(yz)$		
A_g	1	1	1	1	1	1	1	1		x^2, y^2, z^2
B_{1g}	1	1	−1	−1	1	1	−1	−1	R_z	xy
B_{2g}	1	−1	1	−1	1	−1	1	−1	R_y	xz
B_{3g}	1	−1	−1	1	1	−1	−1	1	R_x	yz
A_u	1	1	1	1	−1	−1	−1	−1		
B_{1u}	1	1	−1	−1	−1	−1	1	1	z	
B_{2u}	1	−1	1	−1	−1	1	−1	1	y	
B_{3u}	1	−1	−1	1	−1	1	1	−1	x	

D_{3h}	E	$2C_3$	$3C_2$	σ_h	$2S_3$	$3\sigma_v$		
A_1'	1	1	1	1	1	1		x^2+y^2, z^2
A_2'	1	1	−1	1	1	−1	R_z	
E'	2	−1	0	2	−1	0	(x, y)	(x^2-y^2, xy)
A_1''	1	1	1	−1	−1	−1		
A_2''	1	1	−1	−1	−1	1	z	
E''	2	−1	0	−2	1	0	(R_x, R_y)	(xz, yz)

D_{4h}	E	$2C_4$	C_2	$2C_2'$	$2C_2''$	i	$2S_4$	σ_h	$2\sigma_v$	$2\sigma_d$		
A_{1g}	1	1	1	1	1	1	1	1	1	1		x^2+y^2, z^2
A_{2g}	1	1	1	−1	−1	1	1	1	−1	−1	R_z	
B_{1g}	1	−1	1	1	−1	1	−1	1	1	−1		x^2-y^2
B_{2g}	1	−1	1	−1	1	1	−1	1	−1	1		xy
E_g	2	0	−2	0	0	2	0	−2	0	0	(R_x, R_y)	(xz, yz)
A_{1u}	1	1	1	1	1	−1	−1	−1	−1	−1		
A_{2u}	1	1	1	−1	−1	−1	−1	−1	1	1	z	
B_{1u}	1	−1	1	1	−1	−1	1	−1	−1	1		
B_{2u}	1	−1	1	−1	1	−1	1	−1	1	−1		
E_u	2	0	−2	0	0	−2	0	2	0	0	(x, y)	

D_{5h}	E	$2C_5$	$2C_5^2$	$5C_2$	σ_h	$2S_5$	$2S_5^3$	$5\sigma_v$		
A_1'	1	1	1	1	1	1	1	1		x^2+y^2, z^2
A_2'	1	1	1	−1	1	1	1	−1	R_z	
E_1'	2	2 cos 72°	2 cos 144°	0	2	2 cos 72°	2 cos 144°	0	(x, y)	
E_2'	2	2 cos 144°	2 cos 72°	0	2	2 cos 144°	2 cos 72°	0		(x^2-y^2, xy)
A_1''	1	1	1	1	−1	−1	−1	−1		
A_2''	1	1	1	−1	−1	−1	−1	1	z	
E_1''	2	2 cos 72°	2 cos 144°	0	−2	−2 cos 72°	−2 cos 144°	0	(R_x, R_y)	(xz, yz)
E_2''	2	2 cos 144°	2 cos 72°	0	−2	−2 cos 144°	−2 cos 72°	0		

D_{6h}	E	$2C_6$	$2C_3$	C_2	$3C_2'$	$3C_2''$	i	$2S_3$	$2S_6$	σ_h	$3\sigma_d$	$3\sigma_v$		
A_{1g}	1	1	1	1	1	1	1	1	1	1	1	1		x^2+y^2, z^2
A_{2g}	1	1	1	1	−1	−1	1	1	1	1	−1	−1	R_z	
B_{1g}	1	−1	1	−1	1	−1	1	−1	1	−1	1	−1		
B_{2g}	1	−1	1	−1	−1	1	1	−1	1	−1	−1	1		
E_{1g}	2	1	−1	−2	0	0	2	1	−1	−2	0	0	(R_x, R_y)	(xz, yz)
E_{2g}	2	−1	−1	2	0	0	2	−1	−1	2	0	0		(x^2-y^2, xy)
A_{1u}	1	1	1	1	1	1	−1	−1	−1	−1	−1	−1		
A_{2u}	1	1	1	1	−1	−1	−1	−1	−1	−1	1	1	z	
B_{1u}	1	−1	1	−1	1	−1	−1	1	−1	1	−1	1		
B_{2u}	1	−1	1	−1	−1	1	−1	1	−1	1	1	−1		
E_{1u}	2	1	−1	−2	0	0	−2	−1	1	2	0	0	(x, y)	
E_{2u}	2	−1	−1	2	0	0	−2	1	1	−2	0	0		

7. The D_{nd} Groups ($n = 2, 3, 4, 5, 6$)

$D_{2d} = V_d$	E	$2S_4$	C_2	$2C_2'$	$2\sigma_d$		
A_1	1	1	1	1	1		x^2+y^2, z^2
A_2	1	1	1	−1	−1	R_z	
B_1	1	−1	1	1	−1		x^2-y^2
B_2	1	−1	1	−1	1	z	xy
E	2	0	−2	0	0	(x, y) (R_x, R_y)	(xz, yz)

D_{3d}	E	$2C_3$	$3C_2$	i	$2S_6$	$3\sigma_d$		
A_{1g}	1	1	1	1	1	1		x^2+y^2, z^2
A_{2g}	1	1	−1	1	1	−1	R_z	
E_g	2	−1	0	2	−1	0	(R_x, R_y)	(x^2-y^2, xy) (xz, yz)
A_{1u}	1	1	1	−1	−1	−1		
A_{2u}	1	1	−1	−1	−1	1	z	
E_u	2	−1	0	−2	1	0	(x, y)	

D_{4d}	E	$2S_8$	$2C_4$	$2S_8^3$	C_2	$4C_2'$	$4\sigma_d$		
A_1	1	1	1	1	1	1	1		x^2+y^2, z^2
A_2	1	1	1	1	1	−1	−1	R_z	
B_1	1	−1	1	−1	1	1	−1		
B_2	1	−1	1	−1	1	−1	1	z	
E_1	2	$\sqrt{2}$	0	$-\sqrt{2}$	−2	0	0	(x, y)	
E_2	2	0	−2	0	2	0	0		(x^2-y^2, xy)
E_3	2	$-\sqrt{2}$	0	$\sqrt{2}$	−2	0	0	(R_x, R_y)	(xz, yz)

D_{5d}	E	$2C_5$	$2C_5^2$	$5C_2$	i	$2S_{10}^3$	$2S_{10}$	$5\sigma_d$		
A_{1g}	1	1	1	1	1	1	1	1		x^2+y^2, z^2
A_{2g}	1	1	1	−1	1	1	1	−1	R_z	
E_{1g}	2	2 cos 72°	2 cos 144°	0	2	2 cos 72°	2 cos 144	0	(R_x, R_y)	(xz, yz)
E_{2g}	2	2 cos 144°	2 cos 72°	0	2	2 cos 144°	2 cos 72°	0		(x^2-y^2, xy)
A_{1u}	1	1	1	1	−1	−1	−1	−1		
A_{2u}	1	1	1	−1	−1	−1	−1	1	z	
E_{1u}	2	2 cos 72°	2 cos 144°	0	−2	−2 cos 72°	−2 cos 144°	0	(x, y)	
E_{2u}	2	2 cos 144°	2 cos 72°	0	−2	−2 cos 144°	−2 cos 72°	0		

D_{6d}	E	$2S_{12}$	$2C_6$	$2S_4$	$2C_3$	$2S_{12}^5$	C_2	$6C_2'$	$6\sigma_d$		
A_1	1	1	1	1	1	1	1	1	1		x^2+y^2, z^2
A_2	1	1	1	1	1	1	1	−1	−1	R_z	
B_1	1	−1	1	−1	1	−1	1	1	−1		
B_2	1	−1	1	−1	1	−1	1	−1	1	z	
E_1	2	$\sqrt{3}$	1	0	−1	$-\sqrt{3}$	−2	0	0	(x, y)	
E_2	2	1	−1	−2	−1	1	2	0	0		(x^2-y^2, xy)
E_3	2	0	−2	0	2	0	−2	0	0		
E_4	2	−1	−1	2	−1	−1	2	0	0		
E_5	2	$-\sqrt{3}$	1	0	−1	$\sqrt{3}$	−2	0	0	(R_x, R_y)	(xz, yz)

8. The S_n Groups ($n = 4, 6, 8$)

S_4	E	S_4	C_2	S_4^3		
A	1	1	1	1	R_z	x^2+y^2, z^2
B	1	−1	1	−1	z	x^2-y^2, xy
E	$\begin{cases}1\\1\end{cases}$	$\begin{matrix}i\\-i\end{matrix}$	$\begin{matrix}-1\\-1\end{matrix}$	$\begin{matrix}-i\\i\end{matrix}$	$(x, y)(R_x, R_y)$	(xz, yz)

S_6	E	C_3	C_3^2	i	S_6^5	S_6	$\varepsilon = \exp(2\pi i/3)$	
A_g	1	1	1	1	1	1	R_z	x^2+y^2, z^2
E_g	$\begin{cases}1\\1\end{cases}$	$\begin{matrix}\varepsilon\\\varepsilon^*\end{matrix}$	$\begin{matrix}\varepsilon^*\\\varepsilon\end{matrix}$	$\begin{matrix}1\\1\end{matrix}$	$\begin{matrix}\varepsilon\\\varepsilon^*\end{matrix}$	$\begin{matrix}\varepsilon^*\\\varepsilon\end{matrix}$	(R_x, R_y)	(x^2-y^2, xy) (xz, yz)
A_u	1	1	1	−1	−1	−1	z	
E_u	$\begin{cases}1\\1\end{cases}$	$\begin{matrix}\varepsilon\\\varepsilon^*\end{matrix}$	$\begin{matrix}\varepsilon^*\\\varepsilon\end{matrix}$	$\begin{matrix}-1\\-1\end{matrix}$	$\begin{matrix}-\varepsilon\\-\varepsilon^*\end{matrix}$	$\begin{matrix}-\varepsilon^*\\-\varepsilon\end{matrix}$	(x, y)	

S_8	E	S_8	C_4	S_8^3	C_2	S_8^5	C_4^3	S_8^7	$\varepsilon = \exp(2\pi i/8)$	
A	1	1	1	1	1	1	1	1	R_z	x^2+y^2, z^2
B	1	−1	1	−1	1	−1	1	−1	z	
E_1	$\begin{cases}1\\1\end{cases}$	$\begin{matrix}\varepsilon\\\varepsilon^*\end{matrix}$	$\begin{matrix}i\\-i\end{matrix}$	$\begin{matrix}-\varepsilon^*\\-\varepsilon\end{matrix}$	$\begin{matrix}-1\\-1\end{matrix}$	$\begin{matrix}-\varepsilon\\-\varepsilon^*\end{matrix}$	$\begin{matrix}-i\\i\end{matrix}$	$\begin{matrix}\varepsilon^*\\\varepsilon\end{matrix}$	(x, y) (R_x, R_y)	
E_2	$\begin{cases}1\\1\end{cases}$	$\begin{matrix}i\\-i\end{matrix}$	$\begin{matrix}-1\\-1\end{matrix}$	$\begin{matrix}-i\\i\end{matrix}$	$\begin{matrix}1\\1\end{matrix}$	$\begin{matrix}i\\-i\end{matrix}$	$\begin{matrix}-1\\-1\end{matrix}$	$\begin{matrix}-i\\i\end{matrix}$		(x^2-y^2, xy)
E_3	$\begin{cases}1\\1\end{cases}$	$\begin{matrix}-\varepsilon^*\\-\varepsilon\end{matrix}$	$\begin{matrix}-i\\i\end{matrix}$	$\begin{matrix}\varepsilon\\\varepsilon^*\end{matrix}$	$\begin{matrix}-1\\-1\end{matrix}$	$\begin{matrix}\varepsilon^*\\\varepsilon\end{matrix}$	$\begin{matrix}i\\-i\end{matrix}$	$\begin{matrix}-\varepsilon\\-\varepsilon^*\end{matrix}$		(xz, yz)

9. The Cubic Groups

T	E	$4C_3$	$4C_3^2$	$3C_2$		$\varepsilon = \exp(2\pi i/3)$
A	1	1	1	1		$x^2+y^2+z^2$
E	$\begin{cases}1 \\ 1\end{cases}$	ε ε^*	ε^* ε	$\begin{rcases}1 \\ 1\end{rcases}$		$(x^2-y^2, 2z^2-x^2-y^2)$
T	3	0	0	−1	(x, y, z) (R_x, R_y, R_z)	(xy, xz, yz)

T_d	E	$8C_3$	$3C_2$	$6S_4$	$6\sigma_d$		
A_1	1	1	1	1	1		$x^2+y^2+z^2$
A_2	1	1	1	−1	−1		
E	2	−1	2	0	0		$(2z^2-x^2-y^2, x^2-y^2)$
T_1	3	0	−1	1	−1	(R_x, R_y, R_z)	
T_2	3	0	−1	−1	1	(x, y, z)	(xy, xz, yz)

T_h	E	$4C_3$	$4C_3^2$	$3C_2$	i	$4S_6$	$4S_6^2$	$3\sigma_d$		$\varepsilon = \exp(2\pi i/3)$
A_g	1	1	1	1	1	1	1	1		$x^2+y^2+z^2$
E_g	$\begin{cases}1 \\ 1\end{cases}$	ε ε^*	ε^* ε	1 1	1 1	ε ε^*	ε^* ε	$\begin{rcases}1 \\ 1\end{rcases}$		$(2z^2-x^2-y^2,$ $x^2-y^2)$
T_g	3	0	0	−1	3	0	0	−1	(R_x, R_y, R_z)	(xy, yz, xz)
A_u	1	1	1	1	−1	−1	−1	−1		
E_u	$\begin{cases}1 \\ 1\end{cases}$	ε ε^*	ε^* ε	1 1	−1 −1	$-\varepsilon$ $-\varepsilon^*$	$-\varepsilon^*$ $-\varepsilon$	$\begin{rcases}-1 \\ -1\end{rcases}$		
T_u	3	0	0	−1	−3	0	0	1	(x, y, z)	

O	E	$8C_3$	$3C_2$	$6C_4$	$6C_2'$		
A_1	1	1	1	1	1		$x^2+y^2+z^2$
A_2	1	1	1	−1	−1		
E	2	−1	2	0	0		$(2z^2-x^2-y^2,$ $x^2-y^2)$
T_1	3	0	−1	1	−1	(x, y, z) (R_x, R_y, R_z)	
T_2	3	0	−1	−1	1		(xy, xz, yz)

O_h	E	$8C_3$	$6C_2$	$6C_4$	$3C_2(=C_4^2)$	i	$6S_4$	$8S_6$	$3\sigma_h$	$6\sigma_d$		
A_{1g}	1	1	1	1	1	1	1	1	1	1		$x^2+y^2+z^2$
A_{2g}	1	1	−1	−1	1	1	−1	1	1	−1		
E_g	2	−1	0	0	2	2	0	−1	2	0		$(2z^2-x^2-y^2,$ $x^2-y^2)$
T_{1g}	3	0	−1	1	−1	3	1	0	−1	−1	(R_x, R_y, R_z)	
T_{2g}	3	0	1	−1	−1	3	−1	0	−1	1		(xz, yz, xy)
A_{1u}	1	1	1	1	1	−1	−1	−1	−1	−1		
A_{2u}	1	1	−1	−1	1	−1	1	−1	−1	1		
E_u	2	−1	0	0	2	−2	0	1	−2	0		
T_{1u}	3	0	−1	1	−1	−3	−1	0	1	1	(x, y, z)	
T_{2u}	3	0	1	−1	−1	−3	1	0	1	−1		

10 The Icosahedral Groups I, I_h

I	E	$12C_5$	$12C_5^2$	$20C_3$	$15C_2$		
A	1	1	1	1	1		$x^2+y^2+z^2$
T_1	3	−2 cos 144°	−2 cos 72°	0	−1	(x, y, z) (R_x, R_y, R_z)	
T_2	3	−2 cos 72°	−2 cos 144°	0	−1		
G	4	−1	−1	1	0		
H	5	0	0	−1	1		$(2z^2-x^2-y^2,$ x^2-y^2 $xy, yz, zx)$

I_h	E	$12C_5$	$12C_5^2$	$20C_3$	$15C_2$	i	$12S_{10}$	$12S_{10}^3$	$20S_6$	15σ		
A_g	1	1	1	1	1	1	1	1	1	1		$x^2+y^2+z^2$
T_{1g}	3	$-2\cos 144°$	$-2\cos 72°$	0	−1	3	$-2\cos 72°$	$-2\cos 144°$	0	−1	(R_x, R_y, R_z)	
T_{2g}	3	$-2\cos 72°$	$-2\cos 144°$	0	−1	3	$-2\cos 144°$	$-2\cos 72°$	0	−1		
G_g	4	−1	−1	1	0	4	−1	−1	1	0		
H_g	5	0	0	−1	1	5	0	0	−1	1		$(2z^2-x^2-y^2,$ $x^2-y^2,$ $xy, yz, zx)$
A_u	1	1	1	1	1	−1	−1	−1	−1	−1		
T_{1u}	3	$-2\cos 144°$	$-2\cos 72°$	0	−1	−3	$2\cos 72°$	$2\cos 144°$	0	1	(x, y, z)	
T_{2u}	3	$-2\cos 72°$	$-2\cos 144°$	0	−1	−3	$2\cos 144°$	$2\cos 72°$	0	1		
G_u	4	−1	−1	1	0	−4	1	1	−1	0		
H_u	5	0	0	−1	1	−5	0	0	1	−1		

11. The $C_{\infty v}$ and $D_{\infty h}$ Groups (for linear molecules)

$C_{\infty v}$	E	$2C_\infty^\phi$	$\cdots$	$\infty\sigma_v$		
$A_1 \equiv \Sigma^+$	1	1	$\cdots$	1	z	x^2+y^2, z^2
$A_2 \equiv \Sigma^-$	1	1	$\cdots$	−1	R_z	
$E_1 \equiv \Pi$	2	$2\cos\phi$	$\cdots$	0	$(x, y)(R_x, R_y)$	(xz, yz)
$E_2 \equiv \Delta$	2	$2\cos 2\phi$	$\cdots$	0		(x^2-y^2, xy)
$E_3 \equiv \Phi$	2	$2\cos 3\phi$	$\cdots$	0		
$\cdots$	$\cdots$	$\cdots$	$\cdots$	$\cdots$		

$D_{\infty h}$	E	$2C_\infty^\phi$	$\cdots$	$\infty\sigma_v$	i	$2S_\infty^\phi$	$\cdots$	∞C_2		
$A_{1g} \equiv \Sigma_g^+$	1	1	$\cdots$	1	1	1	$\cdots$	1		x^2+y^2, z^2
$A_{2g} \equiv \Sigma_g^-$	1	1	$\cdots$	−1	1	1	$\cdots$	−1	R_z	
$E_{1g} \equiv \Pi_g$	2	$2\cos\phi$	$\cdots$	0	2	$-2\cos\phi$	$\cdots$	0	(R_x, R_y)	(xz, yz)
$E_{2g} \equiv \Delta_g$	2	$2\cos 2\phi$	$\cdots$	0	2	$2\cos 2\phi$	$\cdots$	0		(x^2-y^2, xy)
$\cdots$	$\cdots$	$\cdots$	$\cdots$	$\cdots$	$\cdots$	$\cdots$	$\cdots$	$\cdots$		
$A_{1u} \equiv \Sigma_u^-$	1	1	$\cdots$	−1	−1	−1	$\cdots$	1		
$A_{2u} \equiv \Sigma_u^+$	1	1	$\cdots$	1	−1	−1	$\cdots$	−1	z	
$E_{1u} \equiv \Pi_u$	2	$2\cos\phi$	$\cdots$	0	−2	$2\cos\phi$	$\cdots$	0	(x, y)	
$E_{2u} \equiv \Delta_u$	2	$2\cos 2\phi$	$\cdots$	0	−2	$-2\cos 2\phi$	$\cdots$	0		
$\cdots$	$\cdots$	$\cdots$	$\cdots$	$\cdots$	$\cdots$	$\cdots$	$\cdots$	$\cdots$		

Appendix D
Reading List

The following brief list represents those books that have been particularly useful to the author in his own study and in his teaching of others. Although they do reflect a particular bias and point of view it is felt they can be generally recommended to the reader for the general improvement of his background and further study.

Additional background as well as introductory material on sets, matrices, vectors, and vector spaces can be obtained from:

J. A. Green, *Sets and Groups*, Dover, New York, 1965.

T. R. Halmos, *Finite-Dimensional Vector Spaces*, 2nd ed., Van Nostrand Reinhold, New York, 1958.

G. Stephenson, *Matrices, Sets and Groups*, American Elsevier, New York, 1965.

The more readable general introductory books on group theory include:

F. A. Cotton, *Chemical Applications of Group Theory*, 2nd ed., Wiley-Interscience, New York, 1971.

L. H. Hall, *Group Theory and Symmetry in Chemistry*, McGraw-Hill, New York, 1969.

R. M. Hochstrasser, *Molecular Aspects of Symmetry*, W. A. Benjamin, New York, 1966.

H. H. Jaffe and M. Orchin, *Symmetry in Chemistry*, Wiley, New York, 1965.

R. McWeeny, *Symmetry*, Macmillan, New York, 1963.

D.S.Schonland, *Molecular Symmetry*, Van Nostrand, London, 1965.

M. Tinkham, *Group Theory and Quantum Mechanics*, McGraw-Hill, New York, 1964.

Somewhat more advanced treatments of group theory and its applications are given by:

M. Hamermesh, *Group Theory*, Addison-Wesley, Reading, Mass., 1962.

L. Jansen and M. Boon, *Theory of Finite Groups. Applications in Physics*, Wiley, New York, 1967.

P. H. E. Meijer and E. Bauer, *Group Theory*, North-Holland, Amsterdam, 1962.

E. P. Wigner, *Group Theory*, Academic, New York, 1959.

In addition, many of the above books contain their own bibliographies which will lead the reader to additional references on both the mathematical aspects of group theory as well as its many applications in science.

Index

Abelian group, 9, 10
Abstract direct product, 25
Accidental degeneracy, 155, 156, 204
Addition of representations, 72
Adjoint, matrix, 59, 148
  operator, 148
Algebraic structure of sets, 5
Alternating group, 50, 184
Antilinear operator, 150
Antisymmetrization, operator, 185
  and renormalization operator, 188
Antisymmetrized orbital products, 187
Antisymmetrized product, 173
Antiunitary operator, 150
Associative law, 4, 7
Atkins, P. W., 233
Axial vector, 124

Band structure, 215
Basis or basis set, 57
Bauer, E., 155, 159, 250
Bloch functions, 205
Block form of a matrix, 72
Boon, M., 155, 159, 249
Bosons, 51, 184

C_{3V} group multiplication table, 15
Cartesian product, 5
Cayley's theorem, 46
Character, complex conjugate, 106
  definition of, 95
  orthogonality relations, 96
  projection operators and, 132
  of a representation, 95
  sum rule, 99
  table, 99
Child, M. S., 233
Class, 26
  algebra, 225
  constants, 226
  geometrical interpretation of, 29
  in variance properties of, 225
  irreducible representations, 98
  order, 226
  in point groups, 29
  product of, 226
  structure of permutations, 51
  transitive property of, 27
  vectors, 229
Closure of a group, 9
Commutation, of operators, 146
  of set elements, 6
Commutative law for sets, 4
Commuting operators, 146, 172
Complement of a set, 3
Complex conjugate representatives, 72
  notation, 101
Complex conjugation operator, 158
Composition law for elements of a set, 5
Configuration, 183
Configuration interaction, 183
Conjugate elements, 26
  and subgroups, 40
Coordinate substitution, 120
Corepresentation, 158, 165
Correlation tables, 232
Cosets, 19
Cotton, F. A., 249
Cross, P. C., 233
Cyclic group, 11, 106

Decius, J. C., 233
Decomposition, of the direct product, 170
  of reducible representations, 101
Degeneracy, 145, accidental, 155, 156, 204
  essential, 156, 204
Degree of a permutation, 45
Dependent product functions, 172
Difference of sets, 3
Dimension of a representation, 63
Dimmock, J. O., 233
Directed valence, 206
Direct factors, 24
Direct product, abstract, 25
  decomposition, 170
  of groups, 24
  within a group, 76
  and irreducible representations, 103

of matrices (Kronecker product), 74
of representations, 74, 168
Direct sum of representations, 73
Disjoint sets, 3
Distributive law for sets, 4
Domain of a mapping, 36
Duality principle of sets, 4

Eigenfunction, 145
Eigenvalues, 145
Elements of sets, 1
Equivalent orbitals, 208
Equivalent representations, 72
Essential degeneracies, 156, 204
Exclusion principle, 187
Eyring, H., 155, 159

Factor groups, 22
Faithful representations, 63
Fermions, 51, 184
Fiber, 36

Generating element, 11
Green, J. A., 249
Group, definition, 19
projection operators, 130
Gruppoids, 9

Hall, L. H., 249
Halmos, T. R., 249
Hamermesh, M., 155, 159, 249
Hermitian matrix, 59
Hermitian operators, 145, 149
Hochstrasser, R. M., 249
Hoffman, R., 200
Homomorphic mapping or homomorphism, 37
Homorphism condition for matrix representation, 66
Hybridization, 206

Idempotent operator, 138
Identity element, 9
mapping, 37
or trivial representation, 64
Image, 36
Improper subgroup, 13
Index of a subgroup, 20
Intersection of sets, 2
Invariance property of classes, 225
Invariant subgroup, 22
Inverse, of a group element, 9
mapping, 36
operator, 148
Irreducibility, criteria, 85, 104
postulate, 155
Irreducible representations, 74, 85
and classes, 98
of direct product groups, 103
Isomorphic groups, 12
Isomorphism or Isomorphic mapping, 37
Isomorphism theorem, 39

Jaffe, H. H., 249
Jansen, L., 155, 159, 249

Kernel of a homomorphic mapping, 39
Kimball, G. E., 155, 159, 215
Koster, G. E., 233

Lagrange's theorem, 19
Linear operators, 150
Linear transformations, 65
Linear vector spaces, 58
Longuet-Higgins, H. C., 152, 159

Mapping, of groups, 37
multiplicity, 36
and permutations, 36
products, 37
range, 36
of sets, 7, 35
and transformations, 36
Master representation, 71, 105
see also regular representation
Matrix representation, 115
McWeeny, R., 249
Meijer, P. H. E., 155, 159, 250
Metric matrix or metric tensor, 59, 148
Molecular orbital theory, 199
Multiplication, of sets or group elements, 5
tables, 10

Normal subgroup, 20
Norm of a vector, 59
Null set, 2

Operator representatives, 147
Orbital product, 182
Orbitals, 180

Orchin, M., 249
Order, of a class, 226
  of an element, 13
  of a group, 9
Ordered pairs of elements, 5
Order structure of sets, 5
Orthogonality, of functions, 61
  relations, 89, 96
  of vectors, 59
Orthonormal bases, 60

Permutation operators, 187
Permutations, 43
  class structure of, 51
  and cycles, 44
  degree of, 45
  and mapping, 36
  regular, 49, 70
Permutation symmetry of wave functions, 180
Phillips, C. S. G., 233
Polar vector, 124
Product, antisymmetrized, 173
  cartesian, 5
  of classes, 226
  of groups, 23
  of mappings, 37
  of sets, 5
  symmetrized, 173
Projection operators, 129, 193
  of groups, 130
  involving characters, 132
Proper subgroup, 13
Proper subsets, 4

Quotient sets, 20

Range of a mapping, 36
Rearrangement theorem, 10
Reducible representations, decomposition, 101
Regular permutations, 49, 70
Regular representation, 70, 71, 104, 105
Representations, of abelian groups, 109
  addition of, 72
  carrier (or representation), space for, 66
  character of, 95
  complex conjugate, 72
  dimension of, 63
  direct product of, 74, 168
  of a direct product group, 77
  direct sum of, 73
  equivalent, 72
  faithful, 63
  identity, 64
  irreducible, 74, 85, 98, 103
  master or regular, 70, 71, 104, 105
  notation for, 100
  or realizations of groups, 63
  unitary, 88

Scalar product, 59, 147
  of functions, 61
Scalar representative, 148
Schönflies notation, 235
Schonland, D. S., 249
Schrödinger equation, 145
Schür's lemmas, 88
  and stable subspaces, 92
Self-adjoint or hermitian matrices, 59
Semigroups, 9
Set structure, 5
Similarity transformation 29, 61
Slater determinant, 186
Spanning of a vector space, 131
sp^n hybrids, 210, 215
Stable subsets, 12
Stable vector spaces, 64
Stationary states, 146
Statz, H., 233
Stephenson, G., 249
Subgroups, 12
  conjugate, 40
  improper or trivial, 13
  index of, 20
  invariant, 22
  normal, 20
  proper, 13
Subsets, 2
Substitution operators, 121
Sum rule, for characters, 99
  for matrix elements, 169
Superdiagonal form of a matrix, 86
Symmetric group, 45, 183
Symmetrized product, 173
Symmetry adapted basis set, 115
Symmetry adapted functions, 137
Symmetry group of the Hamiltonian, 151
Symmetry operator, 151
Symmetry reducing perturbations, 231

Tinkham, M., 249
Transformation, of coordinates, 115
  of functions, 118
  linear, 65
  in mapping, 36
  matrix, 58
  similarity, 29, 61
Transformation properties of product
    wave functions, 76
Transitive property of classes, 27
Transposition, 50
Trivial subgroup, 13

Union of sets, 2
Unitary matrices, 60
Unitary operators, 149
Unitary representation, 88
Unitary vector spaces, 59
Universal set, 3

Valence bond theory, 206
Vector representative, 115, 148
Vector spaces, definition, 57
  linear, 58
  spanning of, 131
  stable, 64
  unitary, 59
Venn diagrams, 2

Walter, J., 155, 159
Wheeler, R. G., 233
Wigner, E. P., 155, 158, 159, 165,
    250
Wilson, E. B., 233